Library of
Davidson College

Discrete Numerical Methods in Physics and Engineering

ACADEMIC PRESS RAPID MANUSCRIPT REPRODUCTION

This is Volume 107 in
MATHEMATICS IN SCIENCE AND ENGINEERING
A series of monographs and textbooks
Edited by RICHARD BELLMAN, *University of Southern California*

The complete listing of books in this series is available from the Publisher upon request.

Discrete Numerical Methods in Physics and Engineering

Donald Greenspan

*Computer Sciences Department
and
Academic Computing Center
University of Wisconsin
Madison, Wisconsin*

1974

ACADEMIC PRESS, INC.
A Subsidiary of Harcourt Brace Jovanovich, Publishers

COPYRIGHT © 1974, BY ACADEMIC PRESS, INC.
ALL RIGHTS RESERVED.
NO PART OF THIS PUBLICATION MAY BE REPRODUCED OR
TRANSMITTED IN ANY FORM OR BY ANY MEANS, ELECTRONIC
OR MECHANICAL, INCLUDING PHOTOCOPY, RECORDING, OR ANY
INFORMATION STORAGE AND RETRIEVAL SYSTEM, WITHOUT
PERMISSION IN WRITING FROM THE PUBLISHER.

ACADEMIC PRESS, INC.
111 Fifth Avenue, New York, New York 10003

United Kingdom Edition published by
ACADEMIC PRESS, INC. (LONDON) LTD.
24/28 Oval Road, London NW1

Library of Congress Cataloging in Publication Data

Greenspan, Donald.
   Discrete numerical methods in physics and engineering.

   (Mathematics in science and engineering, v. 107)
   Bibliography: p.
   1. Numerical analysis. 2. Mathematical physics.
3. Engineering mathematics. I. Title. II. Series.
QA297.G72   515'.625   73-18463
ISBN 0-12-300350-4

AMS (MOS) 1970 Subject Classification: 65-01

PRINTED IN THE UNITED STATES OF AMERICA

Contents

PREFACE ix
ACKNOWLEDGMENTS xi

Chapter I. NUMERICAL SOLUTION OF ALGEBRAIC AND TRANSCENDENTAL SYSTEMS

1.1 Introduction 1
1.2 Matrices and Linear Systems 1
1.3 Gauss Elimination 5
1.4 Tridiagonal Systems 10
1.5 The Generalized Newton's Method 12
1.6 Remarks 22
    Exercises 23

Chapter II. APPROXIMATE SOLUTION OF PROBLEMS FOR ORDINARY DIFFERENTIAL EQUATIONS

2.1 Introduction 26
2.2 Grid Points and Differences 27
2.3 The Method of Taylor Series 29
2.4 Runge-Kutta Methods 32

2.5 The Nonlinear Pendulum 43
2.6 Instability 46
2.7 Periodic Solutions of van der Pol's Equation 52
2.8 Approximate Solution of Boundary Value Problems 57
2.9 Remarks 65
     Exercises 67

Chapter III. NUMERICAL SOLUTION OF ELLIPTIC
               BOUNDARY VALUE PROBLEMS

3.1 Introduction 69
3.2 Boundary Value Problems for the Laplace Equation 73
3.3 Difference Equation Approximation of Laplace's Equation . . . 80
3.4 Interior and Boundary Lattice Points 83
3.5 The Numerical Method 86
3.6 Numerical Solution of the Exterior Dirichlet Problem 91
3.7 Remark on Neumann and Mixed Type Problems 95
3.8 The General Linear Elliptic Equation with Constant Coefficients . . 95
3.9 Extension to Three Dimensions 98
3.10 The Classical Problem of Capacity 102
3.11 Mildly Nonlinear Problems 106
     Exercises 114

Chapter IV. NUMERICAL SOLUTION OF PARABOLIC
               DIFFERENTIAL EQUATIONS

4.1 Introduction 118
4.2 Stability 121
4.3 An Explicit Numerical Method 129
4.4 An Implicit Numerical Method 130
4.5 The Crank-Nicolson Method 134
4.6 Mildly Nonlinear Problems 138
4.7 A Boundary Value Technique 140
     Exercises 151

Chapter V. NUMERICAL SOLUTION OF THE WAVE
              EQUATION

5.1 Introduction 153
5.2 The Cauchy Problem 155

5.3	Stability	160
5.4	An Explicit Method for Initial-Boundary Problems	165
5.5	An Implicit Method for Initial-Boundary Problems	168
5.6	A Second Implicit Method for Initial-Boundary Problems	171
5.7	Mildly Nonlinear Problems	174
5.8	A Boundary Value Technique	175
5.9	Other Methods	182
	Exercises	183

Chapter VI. APPROXIMATE EXTREMIZATION OF FUNCTIONALS

6.1	Introduction	185
6.2	Extremization of Functionals	185
6.3	A Numerical Method	190
6.4	Geodesics	192
6.5	Free Boundary Value Problems	195
6.6	Variational Problems and Partial Differential Equations	197
6.7	The Plateau Problem	198
	Exercises	205

Chapter VII. APPROXIMATE SOLUTION OF FLUID PROBLEMS

7.1	Introduction	208
7.2	A Prototype Liquid Problem	208
7.3	Biharmonic Problems	227
7.4	A Prototype Time Dependent Fluid Problem	228
7.5	A Boundary Value Technique	229
7.6	The Method of Fromm	237
7.7	The Method of Pearson	239
7.8	Remarks on Three Dimensional Problems	239
7.9	Hyperbolic Systems	240
7.10	Initial Value Problems	249
7.11	The Method of Courant, Isaacson, and Rees	250
7.12	The Lax-Wendroff Method	254
7.13	Other Methods	256
	Exercises	257

Chapter VIII. DISCRETE MODEL THEORY

8.1 Introduction 259
8.2 Particles, Time, and Motion 259
8.3 Velocity and Acceleration 261
8.4 The Law of Motion 262
8.5 Damped Motion in a Nonlinear Force Field 262
8.6 Conservation of Energy 264
8.7 Nonlinear String Vibrations 267
    Exercises 274

Appendix A: MATHEMATICS, THE EXACT SCIENCE . . . 276

Appendix B: FORTRAN PROGRAM NAVSTK 286

REFERENCES AND SOURCES FOR FURTHER READING. . 292

ANSWERS TO SELECTED EXERCISES 305

Subject Index 309

Preface

The development of the high-speed digital computer has had, and continues to have, a revolutionary effect on modern applied science. Immediate evidence is available in the form of a large number of computer-generated numerical solutions of fundamental, unsolved systems of mathematical equations. The diversity of fields being affected includes lunar and planetary astrodynamics, wave diffraction, shock waves, laminar flow of liquids, free-surface fluid flow, weather prediction, thermodynamics, elasticity, electrostatic and gravitational potential, optimal control, n-body problems, vibration theory, molecular interaction, quantum theory, and relativistic collapse. Less obviously, there have been natural, qualitative changes in related mathematical models and theories.

This book attempts to develop a broad spectrum of applications that can be formulated as problems in differential equations in the real domain. Existing analytical theories and techniques will be summarized appropriately so that the reader will understand when he should *not* use a computer. For those problems which cannot be solved analytically, we will develop finite difference, computer-oriented numerical methods for approximating solutions. Indeed, if a *computer algorithm* is defined as a finite sequence of computer operations designed to yield an approximate solution of a given mathematical problem, then this book is concerned primarily with the development of computer algorithms. In this connection, it must be understood that the immense power of the modern digital computer lies in its ability to perform arithmetic operations and to store and retrieve numbers with exceptional speed.

In order to develop in the reader the intuition which will enable him to devise sound, economical methods for his own particular problems, heuristic arguments are emphasized throughout. Sources for the precise mathematical foundations are referenced appropriately for the reader with a mathematically oriented background.

Finally, a few words are in order about the emphasis on difference techniques. It is at times possible, of course, to utilize a continuous method of approximation which, by some criterion, is superior to a finite-difference method. Nevertheless, I have never seen an appropriate difference method fail where a continuous method works, and I have seen difference methods work where continuous methods have failed. The latter is especially noticeable in studies of the Navier-Stokes equations. This tremendous breadth of applicability and its inherent structural simplicity are what make difference methods so exceptionally valuable in any direct, numerical approach to problems of applied, scientific interest.

Acknowledgments

For their generous permission to quote freely from my previous book, *Lectures on the Numerical Solution of Linear, Singular, and Nonlinear Differential Equations,* ©1968, I wish to thank Prentice-Hall, Inc., Englewood Cliffs, New Jersey.

Also, since this book is being published by a photo-offset process from an original manuscript, credit for the typing should be given to Patricia Hanson and for the illustrations to Martha Fritz.

CHAPTER I

NUMERICAL SOLUTION OF ALGEBRAIC AND
TRANSCENDENTAL SYSTEMS

1.1 Introduction

The arithmetical operations performed by modern digital computers are exactly those of classical algebra. For this reason, we will be concerned primarily in this book with two basic problems: that of approximating a differential equation by an algebraic or transcendental equation, and that of solving systems of algebraic or transcendental equations. It is to the latter problem that we turn first.

1.2 Matrices and Linear Systems

The general linear algebraic system of n equations in the n unknowns $x_1, x_2, \ldots, x_n$ can be written in the form

(1.1)
$$\begin{aligned} a_{11}x_1 + a_{12}x_2 + a_{13}x_3 + \cdots + a_{1n}x_n &= b_1 \\ a_{21}x_1 + a_{22}x_2 + a_{23}x_3 + \cdots + a_{2n}x_n &= b_2 \\ \vdots \quad \vdots \quad \vdots \quad \vdots \quad \vdots & \\ a_{n1}x_1 + a_{n2}x_2 + a_{n3}x_3 + \cdots + a_{nn}x_n &= b_n \end{aligned}$$

If the matrices x, b and A are defined by

$$(1.2) \qquad x = \begin{pmatrix} x_1 \\ x_2 \\ \vdots \\ x_n \end{pmatrix}, \quad b = \begin{pmatrix} b_1 \\ b_2 \\ \vdots \\ b_n \end{pmatrix}, \quad A = \begin{pmatrix} a_{11} & a_{12} & \cdots & a_{1n} \\ a_{21} & a_{22} & \cdots & a_{2n} \\ \vdots & \vdots & & \vdots \\ a_{n1} & a_{n2} & \cdots & a_{nn} \end{pmatrix}$$

then it follows from the basic laws of matrix operation that system (1.1) can be written compactly as

$$(1.3) \qquad Ax = b.$$

Equivalent forms (1.1) and (1.3) will be used interchangeably, as is convenient.

Let us <u>assume that A is nonsingular</u>, so that for given A and b, x in (1.3) exists and is unique. Indeed, the components of the vector x can be given explicitly by Cramer's rule in terms of determinants. However, if one attempts to evaluate these determinants and thereby find the exact numerical values of $x_1, x_2, \ldots, x_n$, then Cramer's rule, though reasonable for $n = 2, 3$, and 4, becomes readily intractible for increasing values of n, and other methods must be used. Since we will be interested in relatively large values of n, let us, at the outset, introduce several characteristic properties which many applied problems have in common, and which will enable us to solve system (1.1) both quickly and efficiently.

MATRICES AND LINEAR SYSTEMS

Definition 1.1

System (1.1) is said to be diagonally dominant if and only if

$$|a_{ii}| \geq \sum_{\substack{j=1 \\ i \neq j}}^{n} |a_{ij}|, \quad i = 1, 2, \ldots, n,$$

with strict inequality holding for at least one value of i.

Example

The system

$$4x_1 + 2x_2 + 2x_3 = 1$$
$$x_1 - 3x_2 - x_3 = 6$$
$$x_1 + x_2 + 2x_3 = 0$$

is diagonally dominant.

Definition 1.2

System (1.1) is said to be <u>mildly diagonally dominant</u> if and only if

$$|a_{ii}| \geq \max[\,|a_{i1}|, |a_{i2}|, \ldots, |a_{i,i-1}|, |a_{i,i+1}|, \ldots, |a_{in}|\,],$$
$$i = 1, 2, \ldots, n,$$

with strict inequality holding for at least one value of i.

Example

The system

$$5x_1 + 4x_2 - 3x_3 = 0$$
$$x_1 - 3x_2 - x_3 = 0$$
$$x_1 - 4x_2 + 4x_3 = 1$$

is mildly diagonally dominant, but not diagonally dominant.

Definition 1.3

System (1.1) is said to be tridiagonal if and only if all the elements of matrix A are zero except a_{ii}, $a_{j,j+1}$ and $a_{j+1,j}$, where $i = 1, 2, \ldots, n$; $j = 1, 2, \ldots, n-1$.

Example

The system

$$4x_1 + x_2 = 1$$
$$x_1 - 3x_2 + 7x_3 = 0$$
$$x_2 + 3x_3 - x_4 = -1$$
$$x_3 + x_4 - x_5 = 0$$
$$x_4 - 2x_5 = 1$$

is tridiagonal.

In Definition 1.3, the term <u>tridiagonal</u> is appropriate because, in matrix form (1.3), A has the particular representation

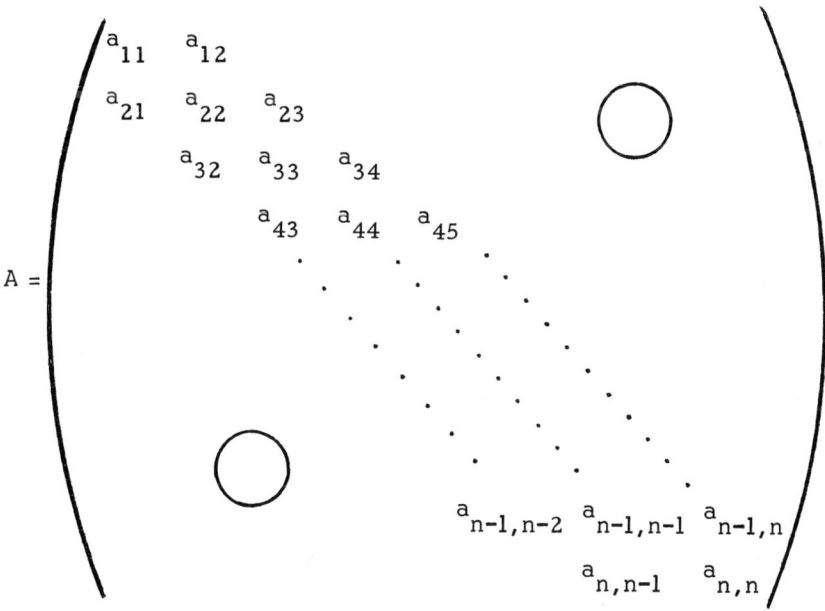

in which all elements are zero except those on the main diagonal and on the diagonals just above and just below the main diagonal.

1.3 Gauss Elimination

In terms of computer capability like that of the UNIVAC 1108, an efficient method for solving many systems when n is relatively small, say $n \leq 400$, is the method of Gauss elimination, which will be described in complete generality after the following illustrative example.

Example

Consider the system

(1.4) $\quad x_1 + 4x_2 - x_3 + x_4 = 2$

(1.5) $\quad x_1 - 2x_2 - 3x_3 + x_4 = 4$

(1.6) $\quad 4x_1 - x_2 + 2x_3 - x_4 = 2$

(1.7) $\quad\quad\quad\;\; x_2 \quad\quad\; - 4x_4 = 0.$

Because, in (1.6), the coefficient of x_1 is in absolute value greater than the absolute value of each of the other coefficients in (1.6), this equation is written separately. Thus,

(1.6) $\quad 4x_1 - x_2 + 2x_3 - x_4 = 2.$

Next, add suitable multiples of (1.6) to each of (1.4), (1.5) and (1.7) to eliminate the x_1 terms in (1.4), (1.5) and (1.7). In this way, (1.4), (1.5) and (1.7) reduce to

(1.4') $\quad \frac{17}{4}x_2 - \frac{3}{2}x_3 + \frac{5}{4}x_4 = \frac{3}{2}$

(1.5') $\quad -\frac{7}{4}x_2 - \frac{7}{2}x_3 + \frac{5}{4}x_4 = \frac{7}{2}$

(1.7') $\quad\quad x_2 \quad\quad\; - 4x_4 = 0.$

Next, because in (1.4') the coefficient of x_2 is in absolute value greater than the absolute value of each of the other coefficients in

GAUSS ELIMINATION

(1.4'), this equation is written separately. Thus

(1.4') $$\frac{17}{4}x_2 - \frac{3}{2}x_3 + \frac{5}{4}x_4 = \frac{3}{2}.$$

Next, add suitable multiples of (1.4') to each of (1.5') and (1.7') to eliminate the x_2 terms in (1.5') and (1.7'). In this way (1.5') and (1.7') reduce to

(1.5") $$-\frac{70}{17}x_3 + \frac{30}{17}x_4 = \frac{70}{17}$$

(1.7") $$\frac{6}{17}x_3 - \frac{73}{17}x_4 = -\frac{6}{17}.$$

Because in (1.5") the coefficient of x_3 is in absolute value greater than the absolute value of each of the other coefficients in (1.5"), this equation is written separately. Thus,

(1.5") $$-\frac{70}{17}x_3 + \frac{30}{17}x_4 = \frac{70}{17}.$$

Next, add a suitable multiple of (1.5") to (1.7") to eliminate the x_3 term in (1.7"). In this way, (1.7") reduces to

(1.7''') $$-\frac{29}{7}x_4 = 0.$$

Thus, system (1.4)-(1.7) has been transformed into the equivalent system (1.6), (1.4'), (1.5"), (1.7'''), that is, to

$$(1.6) \qquad 4x_1 - x_2 + 2x_3 - x_4 = 2$$

$$(1.4') \qquad \frac{17}{4}x_2 - \frac{3}{2}x_3 + \frac{5}{4}x_4 = \frac{3}{2}$$

$$(1.5'') \qquad -\frac{70}{17}x_3 + \frac{30}{17}x_4 = \frac{70}{17}$$

$$(1.7''') \qquad -\frac{29}{7}x_4 = 0 \; .$$

Finally, the latter system is solved by backward substitution, that is, from (1.7''') one has $x_4 = 0$; substitution of $x_4 = 0$ into (1.5'') yields $x_3 = -1$; substitution of $x_4 = 0$, $x_3 = -1$ into (1.4') yields $x_2 = 0$; and substitution of $x_4 = 0$, $x_3 = -1$, $x_2 = 0$ into (1.6) yields $x_1 = 1$, and the original system (1.4)-(1.7) is solved.

Note that, usually, one would simplify an equation like (1.5'') to read

$$-7x_3 + 3x_4 = 7 \; ,$$

if one were working with only pencil and paper. However, this was not done because a digital computer would have divided through and rounded, so that the coefficients would have been finite decimals, and not fractions.

The method illustrated in the above example will now be given a general formulation.

Method of Gauss Elimination

From system (1.1) select an equation in which the coefficient

GAUSS ELIMINATION

of x_1, say a_{k1}, is in absolute value greater than or equal to the absolute value of any other coefficient in the equation. Then, for $j \neq k$, add the multiple $-a_{j1}/a_{k1}$ of the kth equation to the jth equation for each of $j = 1, 2, \ldots, k-2, k-1, k+1, k+2, \ldots, n$. Set the kth equation aside and consider the remaining $(n-1)$ equations, which contain only the $(n-1)$ unknowns $x_2, x_3, \ldots, x_n$. Select from these an equation in which the coefficient of x_2 is in absolute value greater than or equal to the absolute value of any other coefficient in the equation. Add suitable multiples of this equation to the remaining $(n-2)$ equations so that in each resulting equation the x_2 coefficient is zero. Set aside the equation whose x_2 coefficient is non-zero and consider the remaining $(n-2)$ equations in the $(n-2)$ unknowns $x_3, x_4, \ldots, x_n$. In the indicated fashion continue, if possible, the elimination process until, in a finite number of steps, there results a system of equations of the form

$$
\begin{aligned}
c_{11}x_1 + c_{12}x_2 + c_{13}x_3 + \cdots + c_{1,n-1}x_{n-1} + c_{1n}x_n &= C_1 \\
c_{22}x_2 + c_{23}x_3 + \cdots + c_{2,n-1}x_{n-1} + c_{2n}x_n &= C_2 \\
c_{33}x_3 + \cdots + c_{3,n-1}x_{n-1} + c_{3n}x_n &= C_3 \\
\vdots \quad \vdots \\
c_{nn}x_n &= C_n
\end{aligned}
$$

(1.8)

which is equivalent to (1.1), and in which $c_{ii} \neq 0$, $i = 1, 2, \ldots, n$.

Finally, by back-substitution, find $x_n, x_{n-1}, \ldots, x_3, x_2, x_1$ from (1.8).

1.4 Tridiagonal Systems

When system (1.1) is <u>tridiagonal</u> and <u>diagonally dominant</u>, its solution exists and is unique (Geiringer). For such systems, the Gauss elimination method usually can be applied efficiently on a computer like the UNIVAC 1108 for n up to, approximately, 2000, and can be codified precisely as follows. Generate $\beta_1, \beta_2, \ldots, \beta_n$ and $\gamma_1, \gamma_2, \ldots, \gamma_{n-1}$ from

$$(1.9) \quad \beta_1 = a_{11}$$

$$(1.10) \quad \gamma_j = (a_{j,j+1})/\beta_j, \quad j = 1, 2, \ldots, n-1$$

$$(1.11) \quad \beta_j = a_{jj} - a_{j,j-1} \gamma_{j-1}, \quad j = 2, 3, \ldots, n.$$

Next generate $z_1, z_2, \ldots, z_n$ from

$$(1.12) \quad z_1 = b_1/\beta_1$$

$$(1.13) \quad z_j = (b_j - a_{j,j-1} z_{j-1})/\beta_j, \quad j = 2, 3, \ldots, n.$$

Finally, generate the solution $x_1, x_2, \ldots, x_n$ from

$$(1.14) \quad x_n = z_n$$

$$(1.15) \quad x_k = z_k - x_{k+1} \gamma_k, \quad k = n-1, n-2, \ldots, 3, 2, 1.$$

TRIDIAGONAL SYSTEMS

The backward substitution process of the general Gauss elimination procedure is seen clearly from (1.14) and (1.15).

Example

Consider the tridiagonal system

$$-2x_1 + x_2 = 1$$
$$x_1 - 2x_2 + x_3 = 0$$
$$x_2 - 2x_3 + x_4 = 0$$
$$x_3 - 2x_4 + x_5 = 0$$
$$x_4 - 2x_5 = 0 .$$

Then,

(1.16) $\quad a_{11} = a_{22} = a_{33} = a_{44} = a_{55} = -2$

(1.17) $\quad a_{12} = a_{21} = a_{23} = a_{32} = a_{34} = a_{43} = a_{45} = a_{54} = 1$

(1.18) $\quad b_1 = 1, \; b_2 = b_3 = b_4 = b_5 = 0 .$

From (1.9)-(1.11) it follows that

$$\beta_1 = -2$$
$$\gamma_1 = 1/\beta_1 = -\frac{1}{2}$$
$$\beta_2 = -2 - (-\frac{1}{2}) = -\frac{3}{2}$$
$$\gamma_2 = 1/\beta_2 = -\frac{2}{3}$$

$$\beta_3 = -2 - (-\tfrac{2}{3}) = -\tfrac{4}{3}$$
$$\gamma_3 = 1/\beta_3 = -\tfrac{3}{4}$$
$$\beta_4 = -2 - (-\tfrac{3}{4}) = -\tfrac{5}{4}$$
$$\gamma_4 = 1/\beta_4 = -\tfrac{4}{5}$$
$$\beta_5 = -2 - (-\tfrac{4}{5}) = -\tfrac{6}{5} \ .$$

Next, (1.12)-(1.13) yield

$$z_1 = -\tfrac{1}{2}, \ z_2 = -\tfrac{1}{3}, \ z_3 = -\tfrac{1}{4}, \ z_4 = -\tfrac{1}{5}, \ z_5 = -\tfrac{1}{6} \ .$$

Finally, (1.14)-(1.15) imply

$$x_5 = -\tfrac{1}{6}$$
$$x_4 = -\tfrac{1}{5} - (-\tfrac{1}{6})(-\tfrac{4}{5}) = -\tfrac{1}{3}$$
$$x_3 = -\tfrac{1}{4} - (-\tfrac{1}{3})(-\tfrac{3}{4}) = -\tfrac{1}{2}$$
$$x_2 = -\tfrac{1}{3} - (-\tfrac{1}{2})(-\tfrac{2}{3}) = -\tfrac{2}{3}$$
$$x_1 = -\tfrac{1}{2} - (-\tfrac{2}{3})(-\tfrac{1}{2}) = -\tfrac{5}{6} \ .$$

1.5 The Generalized Newton's Method

Consider now an extraordinarily powerful iterative technique for solving important classes of both linear and nonlinear systems called the generalized Newton's method. When applied to linear systems, this method is known in the literature are successive over-relaxation (SOR).

GENERALIZED NEWTON'S METHOD

First, suppose one wishes to determine a real root of a single equation in a single unknown, say

(1.19) $$f(x) = 0$$

where f is continuously differentiable, but not necessarily linear. Let the graph of

(1.20) $$y = f(x)$$

be as shown in Figure 1.1. Of course, the problem of determining the real roots of (1.19) is equivalent to that of finding the real zeros of

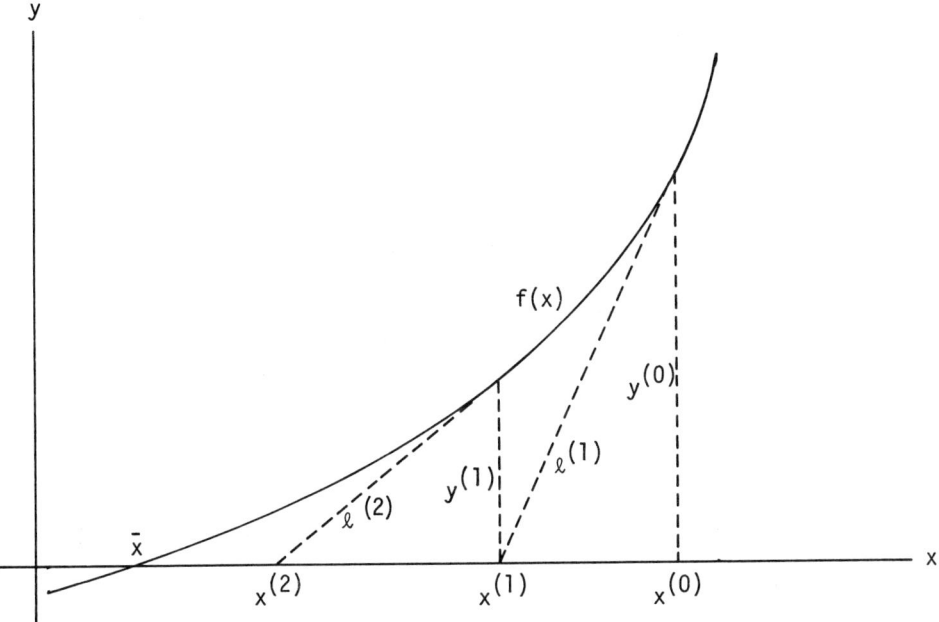

Figure 1.1

(1.20), that is, of finding where the graph of (1.20) meets the X-axis. We shall try to do the latter as follows.

As shown in Figure 1.1, let $\bar{x}$ be a zero of $f(x)$. Since $\bar{x}$ is, in general, unknown, make an initial guess at it, say, $x^{(0)}$. If $f(x^{(0)}) = 0$, then the problem is solved. In all likelihood, however, $f(x^{(0)}) \neq 0$. Then try to get a new approximation, $x^{(1)}$, to $\bar{x}$ which is better than $x^{(0)}$ as follows. Set $y^{(0)} = f(x^{(0)})$. At the point $(x^{(0)}, y^{(0)})$, the slope of the tangent line $\ell^{(1)}$ to $f(x)$ is $f'(x^{(0)})$ and the equation of $\ell^{(1)}$ is

(1.21) $$y - f(x^{(0)}) = (x - x^{(0)}) f'(x^{(0)}) .$$

Let $x^{(1)}$ be the point where $\ell^{(1)}$ meets the X-axis, so that $(x^{(1)}, 0)$ satisfies (1.21). Thus,

(1.22) $$-f(x^{(0)}) = (x^{(1)} - x^{(0)}) f'(x^{(0)}) .$$

Assuming that $\ell^{(1)}$ is not parallel to the X-axis, so that $f'(x^{(0)}) \neq 0$, one has from (1.22) that

(1.23) $$x^{(1)} = x^{(0)} - \frac{f(x^{(0)})}{f'(x^{(0)})} .$$

If $f(x^{(1)}) = 0$, then the problem is solved. If $f(x^{(1)}) \neq 0$, try to improve on approximation $x^{(1)}$ as follows. Set $y^{(1)} = f(x^{(1)})$ and let $\ell^{(2)}$ be the tangent line to $f(x)$ at $(x^{(1)}, y^{(1)})$. If $x^{(2)}$ is the intersection of $\ell^{(2)}$ with the X-axis, then, as in the development of (1.23), it follows that

GENERALIZED NEWTON'S METHOD

(1.24) $$x^{(2)} = x^{(1)} - \frac{f(x^{(1)})}{f'(x^{(1)})}, \quad f'(x^{(1)}) \neq 0.$$

Again, if $f(x^{(2)}) = 0$, then the problem is solved. If $f(x^{(2)}) \neq 0$, construct $x^{(3)}, x^{(4)}, \ldots,$ in the same spirit as $x^{(1)}$ and $x^{(2)}$ were constructed. After $n + 1$ steps, the real number $x^{(n+1)}$ is determined by the formula

(1.25) $$x^{(n+1)} = x^{(n)} - \frac{f(x^{(n)})}{f'(x^{(n)})}, \quad f'(x^{(n)}) \neq 0.$$

The iterative procedure described above is called Newton's method and the recursion formula (1.25) is called Newton's formula.

Under suitable conditions (Ostrowski, Rall (2)), Newton's method can be used to approximate a real root to a very high degree of accuracy.

Of course, it would be of value to have a method which yields a real root in fewer iterations than those of Newton's method. For this reason, instead of constructing the line $\ell^{(1)}$ shown in Figure 1.1, let us try to determine a line through $(x^{(0)}, y^{(0)})$ which intersects the X-axis closer to $\bar{x}$ than $x^{(1)}$. Such a line would have an equation of the form

(1.26) $$y - f(x^{(0)}) = \tau \cdot f'(x^{(0)})(x - x^{(0)}),$$

for the line would differ from $\ell^{(1)}$ only in slope. Setting $y = 0$

and $x = x^{(1)}$ in (1.26) yields

(1.27) $$x^{(1)} = x^{(0)} - \frac{1}{\tau} \frac{f(x^{(0)})}{f'(x^{(0)})}, \quad \tau \cdot f'(x^{(0)}) \neq 0,$$

and just as Newton's formula (1.25) was developed by first considering (1.23), so from (1.27) would follow the recursion formula

(1.28) $$x^{(n+1)} = x^{(n)} - \frac{1}{\tau} \frac{f(x^{(n)})}{f'(x^{(n)})}, \quad \tau \cdot f'(x^{(n)}) \neq 0,$$

For notational simplicity, set $\omega = \frac{1}{\tau}$, so that (1.28) becomes

(1.29) $$x^{(n+1)} = x^{(n)} - \omega \frac{f(x^{(n)})}{f'(x^{(n)})}, \quad f'(x^{(n)}) \neq 0,$$

which is called the generalized Newton's formula.

In (1.29) the constant ω is called an over-relaxation factor, and the modified Newton's method which uses (1.29) in place of (1.25) is called the generalized Newton's method. Of course, Newton's formula results from (1.29) for the special choice $\omega = 1$.

Example

Approximate a positive root of

$$x^3 + \sqrt{3}\, x^2 - 2x = 2\sqrt{3}$$

by the generalized Newton's method.

GENERALIZED NEWTON'S METHOD

Solution

For this simple example, the generalized Newton's formula is

$$(1.30) \qquad x^{(n+1)} = x^{(n)} - \omega \frac{[x^{(n)}]^3 + \sqrt{3}\,[x^{(n)}]^2 - 2x^{(n)} - 2\sqrt{3}}{3[x^{(n)}]^2 + 2\sqrt{3}\,x^{(n)} - 2}.$$

Approximating $\sqrt{3}$ by 1.7, setting $\omega = 1.3$ and $x^{(0)} = 2.0$, and rounding to one decimal place, one has from (1.30) that

$$x^{(1)} = 1.4, \quad x^{(2)} = 1.4.$$

Since $x^{(1)} = x^{(2)}$, further iteration will continue to yield the approximation $x = 1.4$. Repeating the above, but with $\omega = 1$, yields $x^{(1)} = 1.6$, $x^{(2)} = 1.4$, $x^{(3)} = 1.4$, which requires one more step than did the choice $\omega = 1.3$. The exact solution is $x = \sqrt{2}$, which each of the above results approximates correctly to one decimal place.

Suppose next that one has to solve the two equations in two unknowns

$$(1.31) \qquad f_1(x_1, x_2) = 0$$

$$(1.32) \qquad f_2(x_1, x_2) = 0.$$

Then a natural generalization of (1.29) which we shall use for system (1.31)-(1.32) is

(1.33) $$x_1^{(n+1)} = x_1^{(n)} - \omega \frac{f_1(x_1^{(n)}, x_2^{(n)})}{\frac{\partial f_1(x_1^{(n)}, x_2^{(n)})}{\partial x_1}}$$

(1.34) $$x_2^{(n+1)} = x_2^{(n)} - \omega \frac{f_2(x_1^{(n+1)}, x_2^{(n)})}{\frac{\partial f_2(x_1^{(n+1)}, x_2^{(n)})}{\partial x_2}}.$$

It is important to note that in (1.34) the result $x_1^{(n+1)}$, not $x_1^{(n)}$, is used to calculate $x_2^{(n+1)}$. Thus, new data is being utilized as soon as it becomes available.

Example 1

Consider the linear system

$$2x_1 - x_2 = -3$$
$$x_1 - 2x_2 = -3,$$

the exact solution of which is $x_1 = -1$, $x_2 = 1$. Set

$$f_1(x_1, x_2) = 2x_1 - x_2 + 3$$
$$f_2(x_1, x_2) = x_1 - 2x_2 + 3.$$

Then (1.33) and (1.34) take the forms

the application of which, to system (1.41)-(1.45), will be called the generalized Newton's method for systems.

1.6 Remarks

Though the methods presented in this chapter will be adequate for all the applied problems to be considered in future chapters, at times greater economy can be obtained from other methods. In this connection, the interested reader would benefit from an examination of the gradient method, line over-relaxation, Hockney's method, the Peaceman-Rachford method, matrix inversion, the Crout method, general iteration, the square root method, and the method of post-multiplication (see, e.g., Forsythe, Froberg, Goodwin, Hockney, Kunz, Ostrowski, Varga).

It is worth noting that, at present, a good choice for ω in the generalized Newton's method can be determined, or general, only by experimentation in the range $0 < \omega < 2$ (S. Schechter). A choice of ω different from unity often can increase the convergence rate appreciably (Varga).

Finally, note that when one has to solve a system of equations, it is rarely possible to determine, a priori, when an iteration will converge and yield a solution. But, the fact that a certain vector is a solution, no matter how one produced it, can, and should, always be verified by direct substitution into the given system.

GENERALIZED NEWTON'S METHOD

Then the generalized Newton's formulas for (1.41)-(1.45) are

$$(1.46) \quad x_1^{(n+1)} = x_1^{(n)} - \omega \frac{f_1(x_1^{(n)}, x_2^{(n)}, x_3^{(n)}, \ldots, x_{k-1}^{(n)}, x_k^{(n)})}{\dfrac{\partial f_1(x_1^{(n)}, x_2^{(n)}, x_3^{(n)}, \ldots, x_{k-1}^{(n)}, x_k^{(n)})}{\partial x_1}}$$

$$(1.47) \quad x_2^{(n+1)} = x_2^{(n)} - \omega \frac{f_2(x_1^{(n+1)}, x_2^{(n)}, x_3^{(n)}, \ldots, x_{k-1}^{(n)}, x_k^{(n)})}{\dfrac{\partial f_2(x_1^{(n+1)}, x_2^{(n)}, x_3^{(n)}, \ldots, x_{k-1}^{(n)}, x_k^{(n)})}{\partial x_2}}$$

$$(1.48) \quad x_3^{(n+1)} = x_3^{(n)} - \omega \frac{f_3(x_1^{(n+1)}, x_2^{(n+1)}, x_3^{(n)}, \ldots, x_{k-1}^{(n)}, x_k^{(n)})}{\dfrac{\partial f_3(x_1^{(n+1)}, x_2^{(n+1)}, x_3^{(n)}, \ldots, x_{k-1}^{(n)}, x_k^{(n)})}{\partial x_3}}$$

$$\vdots \quad \vdots \quad \vdots$$

$$(1.49) \quad x_{k-1}^{(n+1)} = x_{k-1}^{(n)} - \omega \frac{f_{k-1}(x_1^{(n+1)}, x_2^{(n+1)}, x_3^{(n+1)}, \ldots, x_{k-2}^{(n+1)}, x_{k-1}^{(n)}, x_k^{(n)})}{\dfrac{\partial f_{k-1}(x_1^{(n+1)}, x_2^{(n+1)}, x_3^{(n+1)}, \ldots, x_{k-2}^{(n+1)}, x_{k-1}^{(n)}, x_k^{(n)})}{\partial x_{k-1}}}$$

$$(1.50) \quad x_k^{(n+1)} = x_k^{(n)} - \omega \frac{f_k(x_1^{(n+1)}, x_2^{(n+1)}, \ldots, x_{k-1}^{(n+1)}, x_k^{(n)})}{\dfrac{\partial f_k(x_1^{(n+1)}, x_2^{(n+1)}, \ldots, x_{k-1}^{(n+1)}, x_k^{(n)})}{\partial x_k}},$$

For this system, the generalized Newton's formulas reduce to

(1.37) $$x_1^{(n+1)} = x_1^{(n)} - \omega [e^{x_1^{(n)}} + x_1^{(n)} - 3x_2^{(n)} - 3]/[e^{x_1^{(n)}} + 1]$$

(1.38) $$x_2^{(n+1)} = x_2^{(n)} - \omega [e^{x_2^{(n)}} + x_2^{(n)} - 2x_1^{(n+1)} + 1]/[e^{x_2^{(n)}} + 1].$$

For $x_1^{(0)} = x_2^{(0)} = 0$ and $\omega = 1.5$, (1.37) and (1.38) imply

(1.39) $$x_1^{(1)} = 0 - 1.5[e^0 + 0 - 3 \cdot 0 - 3]/[e^0 + 1] = 1.5$$

(1.40) $$x_2^{(1)} = 0 - 1.5[e^0 + 0 - 2(1.5) + 1]/[e^0 + 1] = 0.75.$$

The results (1.39) and (1.40) would then be inserted into (1.37) and (1.38) to produce $x_1^{(2)}$ and $x_2^{(2)}$, and the iteration would continue in the indicated recursive fashion.

Finally, let us extend (1.33) and (1.34) to the most general system which can occur. Suppose one has to solve the system

(1.41) $$f_1(x_1, x_2, x_3, \ldots, x_{k-1}, x_k) = 0$$

(1.42) $$f_2(x_1, x_2, x_3, \ldots, x_{k-1}, x_k) = 0$$

(1.43) $$f_3(x_1, x_2, x_3, \ldots, x_{k-1}, x_k) = 0$$
$$\vdots$$

(1.44) $$f_{k-1}(x_1, x_2, x_3, \ldots, x_{k-1}, x_k) = 0$$

(1.45) $$f_k(x_1, x_2, x_3, \ldots, x_{k-1}, x_k) = 0.$$

GENERALIZED NEWTON'S METHOD

(1.35) $\qquad x_1^{(n+1)} = x_1^{(n)} - \omega \, [\, 2x_1^{(n)} - x_2^{(n)} + 3]/2$

(1.36) $\qquad x_2^{(n+1)} = x_2^{(n)} - \omega \, [x_1^{(n+1)} - 2x_2^{(n)} + 3]/(-2) \,.$

For initial guess $x_1^{(0)} = x_2^{(0)} = 0$ and for $\omega = 1$, it follows from (1.35) and (1.36) that

$$x_1^{(1)} = -\frac{3}{2}, \; x_2^{(1)} = \frac{3}{4}, \; x_1^{(2)} = -\frac{9}{8}, \; x_2^{(2)} = \frac{15}{16}, \; x_1^{(3)} = -\frac{33}{32},$$

$$x_2^{(3)} = \frac{63}{64}, \; \cdots \,.$$

Thus, for $x_1^{(n+1)}$, one has $x_1^{(0)} = 0$, $x_1^{(1)} = -\frac{3}{2}$, $x_1^{(2)} = -\frac{9}{8}$, $x_1^{(3)} = -\frac{33}{32}$, while, for $x_2^{(n+1)}$, one has $x_2^{(0)} = 0$, $x_2^{(1)} = \frac{3}{4}$, $x_2^{(2)} = \frac{15}{16}$, $x_2^{(3)} = \frac{63}{64}$, which are converging to the correct respective values $x_1 = -1$, $x_2 = 1$.

Example 2

Consider the transcendental system

$$-e^{x_1} - x_1 + 3x_2 + 3 = 0$$

$$e^{x_2} + x_2 - 2x_1 + 1 = 0\,,$$

the exact solution of which is not known. Set

$$f_1(x_1, x_2) = -e^{x_1} - x_1 + 3x_2 + 3$$

$$f_2(x_1, x_2) = e^{x_2} + x_2 - 2x_1 + 1\,.$$

Exercises

1. Show that reordering the equations of system (1.1) may change a nondiagonally dominant system into a diagonally dominant one.

2. Show that every diagonally dominant system is mildly diagonally dominant.

3. Determine which of the following systems are diagonally dominant.

(a)
$$13x_1 + 4x_2 + x_3 - x_4 = 12$$
$$x_1 + 5x_2 + x_3 - x_4 = 0$$
$$2x_1 - x_2 - 6x_3 + 2x_4 = 8$$
$$x_1 - x_2 + x_3 + 7x_4 = 0$$

(b)
$$3x_1 - x_2 = 0$$
$$x_1 - 3x_2 + x_3 = 0$$
$$x_2 - 3x_3 + x_4 = 0$$
$$x_3 - 3x_4 = 0$$

(c)
$$4x_1 - x_2 - x_3 - 3x_4 = 11$$
$$x_1 - 4x_2 + x_3 + x_4 = 0$$
$$x_1 + x_2 - 4x_3 + x_4 = -5$$
$$3x_1 + x_2 + x_3 - 4x_4 = -8$$

(d) $\quad 4x_1 - 3x_2 = 0$

$x_1 - 4x_2 + 3x_3 = 1$

$x_2 - 4x_3 + 3x_4 = 0$

$x_3 - 4x_4 + 3x_5 = 1$

$x_4 - 4x_5 = 0$

(e) $\quad 2x_1 + x_3 + x_5 = 8/15$

$x_1 + 2x_2 + x_4 = 3/4$

$x_1 + x_2 + 2x_3 + x_5 = 7/10$

$x_1 + x_3 + 2x_4 = 7/12$

$x_1 + x_2 + x_4 + 2x_5 = 77/60$

(f) $\quad 4.231x_1 - 0.137x_2 + 0.029x_3 + 0.020x_4 = 3.210$

$-1.031x_1 + 4.397x_2 - 0.332x_3 - 0.115x_4 = -1.001$

$0.415x_1 - 1.447x_2 - 5.137x_3 + 2.014x_4 = 7.394$

$1.974x_1 - 2.106x_2 + 0.847x_3 - 7.130x_4 = -5.214 \ .$

4. Determine which systems in Exericse 3 are mildly diagonally dominant.

5. Determine which systems in Exercise 3 are tridiagonal.

6. If possible, solve each system in Exercise 3 by Gauss elimination. Check your answers.

7. Prove formulas (1.9)–(1.15).

EXERCISES   25

8.   Solve systems (b) and (d) of Exercise 3 by formulas (1.9)–(1.15). Check your answers.

9.   For each of the systems which follow, and for each of the choices $\omega = 1.8,\ 1.4,\ 1.0,\ 0.6,$ and 0.2, find $x_1^{(4)},\ x_2^{(4)}$ and $x_3^{(4)}$ by the generalized Newton's method with $x_1^{(0)} = x_2^{(0)} = x_3^{(0)} = 0$. In each case where the system can be solved exactly, compare the approximate solution with the exact solution and indicate which choice of ω seems most preferable.

(a) $5x_1 + x_2 - 3x_3 = 8$

 $x_1 - 8x_2 + x_3 = 0$

 $3x_1 + 2x_2 - 7x_3 = 0$

(b) $x_1 + x_2 + x_3 = -e^{x_1}$

 $x_1 + x_2 + x_3 = -e^{x_2}$

 $x_1 + x_2 + x_3 = -e^{x_3}$

(c) $2.66x_1 + 1.06x_2 + 1.09x_3 = 0.60$

 $1.06x_1 + 2.66x_2 + 1.09x_3 = 2.26$

 $0.24x_1 + 1.24x_2 + 2.78x_3 = -1.13$.

CHAPTER II

APPROXIMATE SOLUTION OF PROBLEMS FOR ORDINARY
DIFFERENTIAL EQUATIONS

2.1   Introduction

In this chapter attention will be directed to the study of ordinary differential equations of the form

(2.1) $$y'' = f(x,y,y') .$$

Such equations are fundamental in the formulation of dynamical models. The prototype problems associated with (2.1) are the initial value problem, in which one must find a solution of (2.1) for $x \geq a$ which satisfies initial conditions of the form

(2.2) $$y(a) = \alpha, \quad y'(a) = \beta ,$$

and the boundary value problem, in which one must find a solution of (2.1) on $a \leq x \leq b$ which satisfies boundary conditions of the form

(2.3) $$y(a) = \alpha, \quad y(b) = \beta, \quad a < b .$$

In addition to initial and boundary value problems, we will show how to apply initial value techniques to approximate <u>periodic</u> solutions of equations for which auxiliary conditions are <u>not</u> given completely.

The determination of an exact solution of a problem defined by (2.1) and (2.2), or by (2.1) and (2.3), is dependent upon one's ability to construct analytical solutions of (2.1), and very few general analytical methods are available for this purpose. In the case when (2.1) is of the special linear form

$$y'' + P(x)y' + Q(x)y = R(x),$$

then, indeed, solutions can be obtained by means of elementary functions when P and Q are constant and R(x) is of an elementary form (see, e.g., Greenspan (1)), and in terms of series by the method of Frobenius when P(x), Q(x) are rational functions of the form

$$P(x) = \frac{\alpha_1 + \beta_1 x^m}{x(1 + \gamma x^m)}, \quad Q(x) = \frac{\alpha_2 + \beta_2 x^m}{x^2(1 + \gamma x^m)},$$

where α_1, β_1, α_2, β_2, γ, m are constants (see, e.g., Greenspan (1)). If (2.1) is nonlinear, or if it is linear but not of the type described above, then no general analytical method usually exists for constructing solutions, but, indeed, some special trick technique may exist. Such techniques are catalogued, for example, in Kamke. Only after such efforts to find an analytical solution have failed does one then turn to the computer and to numerical methods, which will be studied in the remainder of this chapter.

2.2 Grid Points and Differences

Fundamental to the development of numerical methods for both initial and boundary value problems are the concepts of grid points and differences, which are formulated as follows.

For Δx a positive constant, set $\Delta x = h$ and $x_0 = a$, $x_n = a + n\Delta x$, $n = 1,2,3,\ldots$, where the number of values which n can have may be either finite, or infinite. The symbol $G[a,h]$ will be used consistently to denote the ordered set of points $x_0, x_1, x_2, x_3, x_4, \ldots$, which is called a set of grid points with grid size h. It is only on such point sets that we will consider approximations to a continuous, exact solution $y(x)$ of an initial or boundary value problem. The approximation to $y(x)$ at x_k will be denoted by y_k, $k = 0,1,2,\ldots$.

With regard to difference approximations for first and second order derivatives, the following elementary, but basic, formulas are now recalled:

(2.4) $\quad y'(x) \sim [y(x+h)-y(x)]/h \quad$, (two-point forward formula)

(2.5) $\quad y'(x) \sim [y(x)-y(x-h)]/h \quad$, (two-point backward formula)

(2.6) $\quad y'(x) \sim [y(x+h)-y(x-h)]/(2h) \quad$, (two-point central formula)

(2.7) $\quad y'(x) \sim [-3y(x)+4y(x+h)-y(x+2h)]/(2h)$, (three-point forward formula)

(2.8) $\quad y'(x) \sim [3y(x)-4y(x-h)+y(x-2h)]/(2h) \quad$, (three-point backward formula)

(2.9) $\quad y''(x) \sim [y(x+h)-2y(x)+y(x-h)]/(h^2) \quad$, (three-point central formula).

2.3 The Method of Taylor Series

Let us consider initial value problems first and assume throughout that f, in (2.1), is of such a nature that the solution always exists and is unique.

One of the older methods for approximating a solution of (2.1)-(2.2) is the method of Taylor series. In the past it has not been highly regarded because it requires extensive symbol manipulation in the determination of high-order derivatives. However, since symbol manipulation is now advancing as a computer science discipline, the Taylor series method is returning to a position of stature, and is described as follows.

Let $y(x)$ be the solution of (2.1)-(2.2) and assume that $y(x)$ and $y'(x)$ have Taylor expansions of the form

$$(2.10) \quad y(x+h) = y(x) + hy'(x) + \frac{h^2}{2} y''(x) + \frac{h^3}{3!} y'''(x) + \frac{h^4}{4!} y^{iv}(x) + \cdots$$

$$+ \frac{h^n}{n!} y^{(n)}(x) + \frac{h^{n+1}}{(n+1)!} y^{(n+1)}(\xi), \quad x < \xi < x+h$$

$$(2.11) \quad y'(x+h) = y'(x) + hy''(x) + \frac{h^2}{2} y'''(x) + \frac{h^3}{3!} y^{iv}(x) + \frac{h^4}{4!} y^{v}(x) + \cdots$$

$$+ \frac{h^n}{n!} y^{(n+1)}(x) + \frac{h^{n+1}}{(n+1)!} y^{(n+2)}(\mu), \quad x < \mu < x+h.$$

Let $G[a,h]$ be a set of grid points. Then, from (2.10),

(2.12) $$y(a+h) = y(a) + hy'(a) + \frac{h^2}{2}y''(a) + \frac{h^3}{3!}y'''(a) + \frac{h^4}{4!}y^{iv}(a) + \cdots$$
$$+ \frac{h^n}{n!}y^{(n)}(a) + \frac{h^{n+1}}{(n+1)!}y^{(n+1)}(\xi), \quad a < \xi < a+h.$$

By discarding the remainder term in (2.12) we call

(2.13) $$y_1 = y(a) + hy'(a) + \frac{h^2}{2}y''(a) + \frac{h^3}{3!}y'''(a) + \cdots + \frac{h^n}{n!}y^{(n)}(a)$$

an nth order approximation to $y(a+h)$, and, from (2.11), take

(2.14) $$y_1' = y'(a) + hy''(a) + \frac{h^2}{2}y'''(a) + \frac{h^3}{3!}y^{iv}(a) + \cdots + \frac{h^n}{n!}y^{(n+1)}(a).$$

In (2.13) and (2.14), the values of $y(a)$ and $y'(a)$ are taken from (2.2) while the values of $y''(a), y'''(a), \ldots, y^{(n+1)}(a)$ are obtained from (2.1) and from successive differentiation of (2.1). After constructing y_1 and y_1', one then generates in an analogous fashion y_2 and y_2' from y_1 and y_1', y_3 and y_3' from y_2 and y_2', and so forth, until one terminates the procedure. The nth order Taylor series method may be summarized, then, by the recursion formulas

(2.15) $$y_{k+1} = y_k + hy_k' + \frac{h^2}{2}y_k'' + \frac{h^3}{3!}y_k''' + \cdots + \frac{h^n}{n!}y_k^{(n)}; \quad k = 0,1,\ldots$$

(2.16) $$y_{k+1}' = y_k' + hy_k'' + \frac{h^2}{2}y_k''' + \frac{h^3}{3!}y_k^{iv} + \cdots + \frac{h^n}{n!}y_k^{(n+1)}; \quad k = 0,1,\ldots,$$

where derivatives of order two and higher are obtained from (2.1) and from successive differentiation of (2.1), and where each $y_k^{(i)}$ is assumed to be exact.

TAYLOR SERIES

Example

For $h = 0.3$, find y_0, y_1, and y_2 by means of a third order Taylor series approximation for the initial value problem

(2.17) $$y'' = y^2 - x^2$$

(2.18) $$y(0) = 0, \ y'(0) = 1 .$$

Solution

Since $a = 0$ and $h = 0.3$, the grid points of interest are $a = x_0 = 0$, $x_1 = 0.3$, $x_2 = 0.6$. From (2.18)

(2.19) $$y_0 = 0, \ y'_0 = 1.$$

Since third-order Taylor series approximations are sought, (2.15) and (2.16) take the forms

(2.20) $$y_{k+1} = y_k + (0.3)y'_k + (0.045)y''_k + (0.0045)y'''_k; \ k = 0,1$$

(2.21) $$y'_{k+1} = y'_k + (0.3)y''_k + (0.045)y'''_k + (0.0045)y^{iv}_k; \ k = 0,1 .$$

Also, from (2.17),

(2.22) $$y'' = y^2 - x^2$$

(2.23) $$y''' = 2yy' - 2x$$

(2.24) $$y^{iv} = 2(y')^2 + 2yy'' - 2 .$$

Now, for $k = 0$, one has from (2.19) and (2.22)-(2.24) that

$$y_0 = 0, \; y'_0 = 1, \; y''_0 = y_0^2 - x_0^2 = 0, \; y'''_0 = 2y_0 y'_0 - 2x_0 = 0,$$

$$y_0^{iv} = 2(y'_0)^2 + 2y_0 y''_0 - 2 = 0,$$

which, upon substitution into (2.20) and (2.21), implies

(2.25) $\quad\quad\quad y_1 = 0.3, \; y'_1 = 1$.

Next, for $k = 1$, one has from (2.22)-(2.25) that

$$y_1 = 0.3, \; y'_1 = 1, \; y''_1 = y_1^2 - x_1^2 = 0, \; y'''_1 = 2y_1 y'_1 - 2x_1 = 0,$$

$$y_1^{iv} = 2(y'_1)^2 + 2y_1 y''_1 - 2 = 0,$$

which, upon substitution into (2.20) and (2.21) yields

(2.26) $\quad\quad\quad y_2 = 0.6, \; y'_2 = 1,$

and the example is complete. Note that the numerical solution suggests that that exact solution is $y = x$.

2.4 Runge-Kutta Methods

Runge-Kutta methods are, perhaps, the most popular of the numerical methods for initial value problems because of their simplicity, relatively high accuracy, and broad applicability. It is only in such special cases as when <u>high</u> accuracy is required from <u>large</u> grid sizes, as is the case in problems in astromechanics, that existing

RUNGE-KUTTA METHODS

Runge-Kutta techniques may have little value. These methods do not require the symbol manipulations inherent in the method of Taylor series, but instead make use of values of $f(x,y,y')$ for x not in $G[a,h]$ to yield approximations at points of $G[a,h]$. Of course, since $f(x,y,y')$ is known for many other values of x, it is reasonable not to discard such a large amount of available data.

For intuitive reasons, we will discuss Runge-Kutta methods first for the simple first order equation

(2.27) $$y' = f(x,y) ,$$

and then show how to extend the results in a completely natural way to (2.1).

The following simple method which can be used to approximate a solution of the initial value problem defined by (2.27) and

(2.28) $$y(a) = \alpha$$

was devised by Euler in 1768. On $G[a,h]$, generate y_k; $k = 1, 2, \ldots$ from

(2.29) $$y_0 = \alpha$$

(2.30) $$y_{k+1} = y_k + h f(x_k, y_k), \quad k = 1, 2, \ldots ,$$

where (2.30) is derived from the approximation

$$(2.31) \qquad \frac{y_{k+1} - y_k}{h} = f(x_k, y_k)$$

of (2.27). Though Euler's technique has a firm mathematical basis, it is relatively uneconomical in the sense that to obtain a desired accuracy, one must choose h much smaller than is necessary in other methods. To improve on (2.31), consider the more general formula

$$(2.32) \qquad \frac{y_{k+1} - y_k}{h} = \beta_0 f(x_k, y_k) + \beta_1 f(x_k + \gamma h, y_k + \delta h),$$

where β_0, β_1, γ and δ are parameters to be determined. Note that should γ have a value like 1/2, then $x_k + \gamma h$ would *not* be a grid point, and such a possibility now exists because of the form of (2.32).

In equivalent form, (2.32) can be rewritten as

$$(2.33) \qquad y_{k+1} = y_k + h\beta_0 f(x_k, y_k) + h\beta_1 f(x_k + \gamma h, y_k + \delta h).$$

If f can be written as a Taylor expansion in two variables, say, as

$$f(x_k + \gamma h, y_k + \delta h) = f(x_k, y_k) + h\gamma f_x(x_k, y_k) + h\delta f_y(x_k, y_k)$$
$$+ \frac{1}{2}[h^2\gamma^2 f_{xx}(x_k, y_k) + 2h^2\gamma\delta f_{xy}(x_k, y_k) + h^2\delta^2 f_{yy}(x_k, y_k)] + O(h^3),$$

then, substitution into (2.33) and recombination, implies that

RUNGE-KUTTA METHODS

$$(2.34) \quad y_{k+1} = y_k + h[(\beta_0 + \beta_1)f(x_k, y_k)] + \frac{h^2}{2}[2\beta_1 \gamma f_x(x_k, y_k)$$

$$+ 2\beta_1 \delta f_y(x_k, y_k)] + \frac{h^3}{6}[3\beta_1 \gamma^2 f_{xx}(x_k, y_k)$$

$$+ 6\beta_1 \gamma \delta f_{xy}(x_k, y_k) + 3\beta_1 \delta^2 f_{yy}(x_k, y_k)] + \beta_1 O(h^4).$$

If the exact solution of (2.27) at x_{k+1} can be written as the Taylor series

$$(2.35) \quad y(x_{k+1}) = y(x_k) + hy'(x_k) + \frac{h^2}{2}y''(x_k) + \frac{h^3}{3!}y'''(x_k) + \cdots ,$$

then, from (2.27),

$$(2.36) \quad y'(x_k) = f(x_k, y_k)$$

$$(2.37) \quad y''(x_k) = f_x(x_k, y_k) + f_y(x_k, y_k) f(x_k, y_k)$$

$$(2.38) \quad y'''(x_k) = f_{xx}(x_k, y_k) + 2f(x_k, y_k) f_{xy}(x_k, y_k)$$

$$+ [f(x_k, y_k)]^2 f_{yy}(x_k, y_k) + f_x(x_k, y_k) f_y(x_k, y_k)$$

$$+ f(x_k, y_k)[f_y(x_k, y_k)]^2 ,$$

and substitution of (2.36)-(2.38) into (2.35) yields

$$(2.39) \quad y(x_{k+1}) = y(x_k) + hf(x_k, y_k) + \frac{h^2}{2}[f_x(x_k, y_k) + f_y(x_k, y_k) f(x_k, y_k)]$$

$$+ \frac{h^3}{6}\{f_{xx}(x_k, y_k) + 2f(x_k, y_k) f_{xy}(x_k, y_k) + f_{yy}(x_k, y_k)[f(x_k, y_k)]^2$$

$$+ f_x(x_k, y_k) f_y(x_k, y_k) + f(x_k, y_k)[f_y(x_k, y_k)]^2\} + O(h^4) .$$

Consider now how to choose parameter values β_0, β_1, γ, δ in (2.34) so that the numerical approximation y_{k+1} and the exact value $y(x_{k+1})$ agree in their Taylor expansions through <u>as many</u> terms as possible. If we wish to construct a formula which agrees through the h^2 terms, at least, then, assuming $y_k = y(x_k)$, one finds by setting corresponding coefficients of (2.34) and (2.39) equal that

(2.40) $\qquad \beta_0 + \beta_1 = 1, \quad 2\beta_1 \gamma = 1, \quad 2\beta_1 \delta = f(x_k, y_k)$.

System (2.40) has an infinite number of solutions, one of which is

(2.41) $\qquad \beta_0 = \beta_1 = \frac{1}{2}, \quad \gamma = 1, \quad \delta = f(x_k, y_k)$,

and a second of which is

(2.42) $\qquad \beta_0 = 0, \quad \beta_1 = 1, \quad \gamma = \frac{1}{2}, \quad \delta = \frac{1}{2} f(x_k, y_k)$.

Substitution of (2.41) and (2.42) into (2.33) yields, respectively, the Runge-Kutta formulas

(2.43) $\qquad y_{k+1} = y_k + \frac{h}{2} f(x_k, y_k) + \frac{h}{2} f(x_k + h, y_k + h f(x_k, y_k))$

(2.44) $\qquad y_{k+1} = y_k + h f(x_k + \frac{h}{2}, y_k + \frac{h}{2} f(x_k, y_k))$.

Formulas (2.43) and (2.44) are called second order approximations because they agree with Taylor expansion (2.35) through terms of order h^2.

RUNGE-KUTTA METHODS

For convenience in programming, Runge-Kutta formulas are usually written in the following form, as illustrated for (2.43):

(2.45)
$$\begin{cases} y_{k+1} = y_k + \frac{1}{2}(m_0 + m_1) \\ m_0 = hf(x_k, y_k) \\ m_1 = hf(x_k + h, y_k + m_0). \end{cases}$$

Example

Consider the initial value problem

$$y' = x - y, \quad y(0) = 1.$$

Then (2.45) has the particular form

(2.46)
$$\begin{cases} y_{k+1} = y_k + \frac{1}{2}(m_0 + m_1) \\ m_0 = h(x_k - y_k) \\ m_1 = h(x_k + h - y_k - m_0). \end{cases}$$

For $h = 0.1$, $y_0 = 1$, the numerical solution generated by (2.46) for $k = 0, 1, \ldots, 9$ is

$y_0 = 1.000$, $y_1 = 0.910$, $y_2 = 0.838$, $y_3 = 0.782$,

$y_4 = 0.742$, $y_5 = 0.714$, $y_6 = 0.699$, $y_7 = 0.694$,

$y_8 = 0.700$, $y_9 = 0.714$, $y_{10} = 0.737$.

In general, a Runge-Kutta approximation of

(2.47) $\qquad y' = f(x,y), \quad x_i \leq x \leq x_{i+1}$

is a formula of the form

(2.48) $\quad y_{i+1} = y_i + h[\alpha_1 f(\xi_1,\eta_1) + \alpha_2 f(\xi_2,\eta_2) + \cdots + \alpha_k f(\xi_k,\eta_k)],$

where $x_i \leq \xi_j \leq x_{i+1}$, $j = 1,2,\ldots,k$. Of course, (2.30), (2.43), and (2.44) are special cases of (2.48). A large number of other Runge-Kutta formulas have been developed. For example, Heun showed that

(2.49) $\begin{cases} y_{i+1} = y_i + \dfrac{1}{4}(k_0 + 3k_2) \\[4pt] k_0 = hf(x_i, y_i) \\[4pt] k_1 = hf(x_i + \dfrac{h}{3}, y_i + \dfrac{k_0}{3}) \\[4pt] k_2 = hf(x_i + \dfrac{2h}{3}, y_i + \dfrac{2k_1}{3}) \end{cases}$

approximates (2.35) through terms of order h^3, while Kutta showed that

(2.50) $\begin{cases} y_{i+1} = y_i + \dfrac{1}{6}(k_0 + 2k_1 + 2k_2 + k_3) \\[4pt] k_0 = hf(x_i, y_i) \\[4pt] k_1 = hf(x_i + \dfrac{h}{2}, y_i + \dfrac{k_0}{2}) \\[4pt] k_2 = hf(x_i + \dfrac{h}{2}, y_i + \dfrac{k_1}{2}) \\[4pt] k_3 = hf(x_{i+1}, y_i + k_2) \end{cases}$

RUNGE-KUTTA METHODS

approximates (2.35) through terms of order h^4. For additional formulas of order h^5, h^6, h^7 and h^8, see Fehlberg, Lawson, Luther and Konen, and Sarafyan. At present, no Runge-Kutta formula of order h^9 is known, while the correctness of certain $O(h^8)$ formulas is difficult to verify.

Because Kutta's formula (2.50) provides an excellent balance of simplicity, high-order accuracy and economy, we will concentrate on it. However, the development to follow can be extended to any of the various Runge-Kutta formulas available.

Consider again (2.1), that is

(2.51) $$y'' = f(x,y,y') .$$

In (2.51), set $y' = v$ so that (2.51) is equivalent to the system

(2.52) $$\begin{cases} y' = v \\ v' = f(x,y,v) . \end{cases}$$

But (2.52) is only a particular form of the more general system

(2.53) $$\begin{cases} \dfrac{dy}{dx} = F(x,y,v) \\ \dfrac{dv}{dx} = G(x,y,v) , \end{cases}$$

with

(2.54) $$F(x,y,v) \equiv v, \quad G(x,y,v) \equiv f(x,y,v) .$$

Library of Davidson College

Now, just as (2.50) is a fourth order approximation of (2.47), so it follows in the same way that

(2.55)
$$\begin{cases} y_{i+1} = y_i + \frac{1}{6}(k_0 + 2k_1 + 2k_2 + k_3) \\ v_{i+1} = v_i + \frac{1}{6}(m_0 + 2m_1 + 2m_2 + m_3) \end{cases},$$

where

$$k_0 = hF(x_i, y_i, v_i) \qquad m_0 = hG(x_i, y_i, v_i)$$
$$k_1 = hF(x_i + \frac{h}{2}, y_i + \frac{k_0}{2}, v_i + \frac{m_0}{2}), \quad m_1 = hG(x_i + \frac{h}{2}, y_i + \frac{k_0}{2}, v_i + \frac{m_0}{2})$$
$$k_2 = hF(x_i + \frac{h}{2}, y_i + \frac{k_1}{2}, v_i + \frac{m_1}{2}), \quad m_2 = hG(x_i + \frac{h}{2}, y_i + \frac{k_1}{2}, v_i + \frac{m_1}{2})$$
$$k_3 = hF(x_{i+1}, y_i + k_2, v_i + m_2), \qquad m_3 = hG(x_{i+1}, y_i + k_2, v_i + m_2)$$

are the Kutta formulas for system (2.53). Finally, from (2.54) it follows that the Kutta formulas for system (2.52) are

(2.56)
$$\begin{cases} y_{i+1} = y_i + hv_i + \frac{h}{6}(m_0 + m_1 + m_2) \\ v_{i+1} = v_i + \frac{1}{6}(m_0 + 2m_1 + 2m_2 + m_3) \end{cases}$$

where

(2.57) $$m_0 = hf(x_i, y_i, v_i)$$

(2.58) $$m_1 = hf(x_i + \frac{h}{2}, y_i + \frac{v_i h}{2}, v_i + \frac{m_0}{2})$$

(2.59) $$m_2 = hf(x_i + \frac{h}{2}, y_i + \frac{v_i h}{2} + \frac{m_0 h}{4}, v_i + \frac{m_1}{2})$$

(2.60) $$m_3 = hf(x_{i+1}, y_i + v_i h + \frac{m_1 h}{2}, v_i + m_2).$$

RUNGE-KUTTA METHODS

For initial value problem (2.1)-(2.2), one knows, from (2.2) and (2.52), both y_0 and v_0. These are used first in (2.57)-(2.60) to determine m_0, m_1, m_2, m_3, which are used in turn to generate y_1 and v_1 from (2.56). Next, substitution of y_1 and v_1 into (2.57)-(2.60) generates new values m_0, m_1, m_2, m_3, which are used to generate y_2 and v_2 from (2.56). In the indicated recursive fashion, the process continues and one generates $y_3, y_4, \ldots, y_n$ and $v_3, v_4, \ldots, v_n$.

Example

Consider the initial value problem

$$y'' + y' + y = 1 + x, \quad y(0) = 0, \; y'(0) = 1 ,$$

which, in system form, can be reformulated as

$$y' = v$$
$$v' = 1 + x - y - v$$
$$y(0) = 0, \; v(0) = 1 .$$

Thus,

$$x_0 = 0, \; y_0 = 0, \; v_0 = 1$$

and

$$f(x, y, v) = 1 + x - y - v.$$

For $h = 0.5$, let us show how to generate y_1, v_1, y_2 and v_2 by (2.56)-(2.60). First, then, let $i = 0$ in (2.57)-(2.60), so that

$$m_0 = (0.5) f(0,0,1) = 0$$
$$m_1 = (0.5) f(\tfrac{1}{4}, \tfrac{1}{4}, 1) = 0$$
$$m_2 = (0.5) f(\tfrac{1}{4}, \tfrac{1}{4}, 1) = 0$$
$$m_3 = (0.5) f(\tfrac{1}{2}, \tfrac{1}{2}, 1) = 0 .$$

From (2.56), it follows that

$$y_1 = 0 + (\tfrac{1}{2})(1) + \tfrac{1}{12}(0+0+0) = \tfrac{1}{2}$$
$$v_1 = 1 + \tfrac{1}{6}(0 + 2\cdot 0 + 2\cdot 0 + 0) = 1 .$$

Next, let $i = 2$ in (2.57)-(2.60), so that

$$m_0 = (0.5) f(\tfrac{1}{2}, \tfrac{1}{2}, 1) = 0$$
$$m_1 = (0.5) f(\tfrac{3}{4}, \tfrac{3}{4}, 1) = 0$$
$$m_2 = (0.5) f(\tfrac{3}{4}, \tfrac{3}{4}, 1) = 0$$
$$m_3 = (0.5) f(1, 1, 1) = 0 .$$

From (2.56), it follows finally that

$$y_2 = \tfrac{1}{2} + \tfrac{1}{2}(1) + \tfrac{1}{12}(0+0+0) = 1$$
$$v_2 = 1 + \tfrac{1}{6}(0 + 2\cdot 0 + 2\cdot 0 + 0) = 1 .$$

Note, incidentally, that the approximations y_1 and y_2 coincide precisely, at the grid points, with the solution $y = x$ of the given, rather trivial, initial value problem only because second and higher order derivatives of this exact, continuous solution are all identically zero.

2.5 The Nonlinear Pendulum

Let us consider next an applied, nontrivial problem of long-standing interest and show how to solve it numerically by means of (2.56)-(2.60).

Consider a pendulum, as drawn in Figure 2.1, which has mass m centered at P and is hinged at O. Assume that P is constrained to move on a circle whose radius is ℓ and whose center is O. Let θ be the angular measure, in radians, of the pendulum deviation from the vertical. The problem is that of describing the motion of P after its release from an initial position of rest.

It is known from laboratory experimentation that the motion of the pendulum is damped and that the length of time between consecutive swings decreases.

Analytically, we reason as follows. Assume that the motion of P is determined by Newton's dynamical equation

$$(2.61) \qquad F = ma.$$

Circular arc NP has length $\ell\theta$, so that $a = \dfrac{d^2}{dt^2}(\ell\theta) = \ell\ddot{\theta}$. Thus, (2.61) becomes

(2.62) $$F = m\ell\ddot{\theta}.$$

In considering the force F which acts on P, let F_1 be the gravitational component, so that

(2.63) $$F_1 = -mg\sin\theta, \ g > 0,$$

let F_2 be a damping component of the form

(2.64) $$F_2 = -c\dot{\theta}, \ c > 0, \ c \ \text{a constant},$$

and assume that these are the only forces whose effects are significant. Then,

$$F = -mg\sin\theta - c\dot{\theta},$$

so that (2.62) readily reduces to

(2.65) $$\ddot{\theta} + \frac{c}{m\ell}\dot{\theta} + \frac{g}{\ell}\sin\theta = 0.$$

The problem, then, is one of solving (2.65) subject to the given initial conditions

(2.66) $$\theta(0) = \alpha, \ \dot{\theta}(0) = 0.$$

NONLINEAR PENDULUM

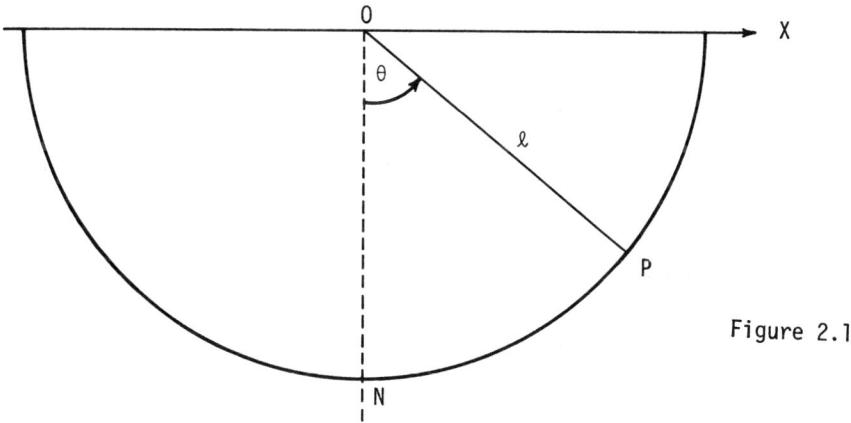

Figure 2.1

For illustrative purposes, let us consider the strongly damped pendulum motion defined by

(2.67) $\quad \ddot{\theta} + (0.3)\dot{\theta} + \sin \theta = 0$

(2.68) $\quad \theta(0) = \pi/4, \ \dot{\theta}(0) = 0.$

No analytical method is known for constructing the exact solution of this problem. Numerically, then, set $t = x$ and $\theta = y$ and solve (2.56)-(2.60) with $\Delta t = h = 0.01$. The computation was carried out in double precision on the UNIVAC 1108 for 15000 time steps, that is, for 150 seconds of pendulum motion, with a total computing time of two seconds. The first 15.0 seconds of pendulum oscillation is shown in Figure 2.2, where the peak, or extreme, values 0.75462, -0.47647, 0.29335, -0.18156, 0.11259, occur at the times 0, 3.28, 6.49, 9.68, 12.86, respectively. The time required for the pendulum to travel from one peak to another decreased monotonically and damping was present during the entire 150 seconds of motion.

Any attempt to linearize (2.67) results in a solution which either does not damp out, or has a constant time interval between successive swings, or both (see e.g., Greenspan (13)).

2.6 Instability

From the computing point of view, a set of calculations which results in overflow is called <u>unstable</u>. Mathematically, there are several formulations of the concept of instability. When instability occurs, one must check first for machine and program errors. If neither is present, one must then try to apply mathematical analysis to the algorithm to find the source, or sources, of the trouble. Two of the more common mathematical sources of instability will be explored now by means of illustrative examples.

Consider, first, the initial value problem

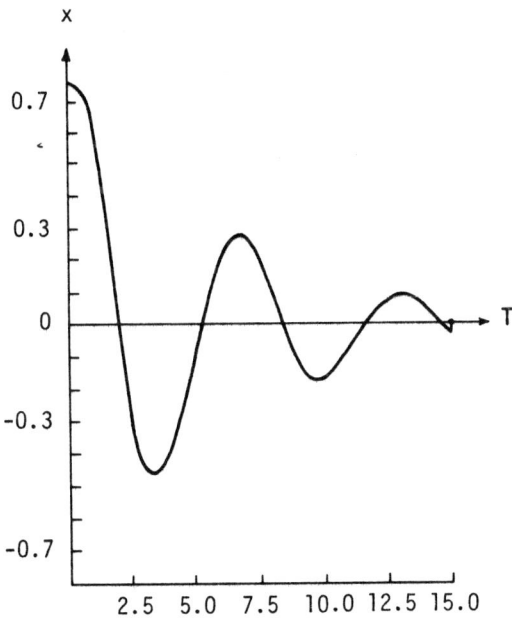

Figure 2.2

(2.69) $$y' = -10y, \quad y(0) = 1,$$

the exact solution of which is

(2.70) $$y = e^{-10x}.$$

This solution tends to zero as x goes to infinity and is always positive. Suppose that one applies Euler's method with $\Delta x = 1.0$ to approximate the solution of (2.69). Then

(2.71) $$y_{i+1} = -9y_i, \quad i = 0, 1, 2, \ldots$$

(2.72) $$y_0 = 1,$$

from which it follows that

$$y_1 = -9, \; y_2 = (-9)^2, \; y_3 = (-9)^3, \; y_4 = (-9)^4, \ldots, \; y_n = (-9)^n, \ldots.$$

Rather quickly, this iteration results in overflow. To analyze the instability, let us proceed as follows. For arbitrary Δx, Euler's method, applied to (2.69), yields

(2.73) $$y_{i+1} = y_i - 10(\Delta x)y_i = y_i(1 - 10\Delta x).$$

Thus,

(2.74) $$\begin{cases} y_1 = y_0(1 - 10\Delta x) = (1 - 10\Delta x) \\ y_2 = y_1(1 - 10\Delta x) = (1 - 10\Delta x)^2 \\ y_3 = (1 - 10\Delta x)^3 \\ \vdots \\ y_n = (1 - 10\Delta x)^n. \end{cases}$$

For instability, the sequence $|y_n|$, $n = 0, 1, 2, \ldots$, must exceed the largest number in one's computer, which, assuming that all computations are exact, can happen only if $|1 - 10\Delta x| > 1$. Thus, <u>to avoid instability</u>, Δx must be chosen to satisfy $|1 - 10\Delta x| \leq 1$, or, $\Delta x \leq 0.2$. Thus, for example, for $\Delta x = 0.01$, Euler's formula for

(2.75) becomes

INSTABILITY

(2.75) $\qquad y_{i+1} = (0.9)y_i, \quad i = 0,1,2,\ldots,$

and iteration with $y_0 = 1$ yields the bounded sequence

(2.76) $\quad y_1 = (0.9), \ y_2 = (0.9)^2, \ y_3 = (0.9)^3, \ y_4 = (0.9)^4, \ldots, y_n = (0.9)^n,$

which, like (2.70), is positive and decreasing.

Now, in the above example it was assumed that computations were exact. Let us show next how computations which are inexact can also be a source of instability. Suppose one is given the difference equation

(2.77) $\qquad 2y_{i+2} - 5y_{i+1} + y_i = 0, \quad i = 0,1,2,\ldots,$

with initial data

(2.78) $\qquad y_0 = 0.5, \quad y_1 = 0.25,$

and is asked to generate $y_2, y_3, y_4, \ldots$. Then, from (2.77) and (2.78),

(2.79) $\qquad y_{i+2} = \frac{5}{2} y_{i+1} - y_i, \quad i = 0,1,2,\ldots,$

and

$$y_2 = \frac{1}{8}, \ y_3 = \frac{1}{16}, \ y_4 = \frac{1}{32}, \ldots, y_n = \left(\frac{1}{2}\right)^{n+1},$$

from which it follows that $y_n \to 0$ as $n \to \infty$. However, suppose that in place of doing exact calculations, as above, one duplicates what a digital computer does and allows roundoff error to be introduced in the following simple way: round off all given data and results of all arithmetic operations to one decimal place. Thus, initially one has

(2.80) $$y_0 = 0.5$$

(2.81) $$y_1 = 0.3,$$

and, from (2.77),

$$y_2 = \frac{5}{2}(0.3) - (0.5) = (2.5)(0.3) - (0.5) = (0.8) - (0.5) = 0.3$$

$$y_3 = \frac{5}{2}(0.3) - (0.3) = 0.5$$

$$y_4 = \frac{5}{2}(0.5) - (0.3) = 1.0$$

$$y_5 = \frac{5}{2}(1.0) - (0.5) = 2.0$$

$$y_6 = \frac{5}{2}(2.0) - (1.0) = 4.0$$

$$y_7 = \frac{5}{2}(4.0) - (2.0) = 8.0$$

$$\vdots$$

$$y_n = 2^{n-4}, \ n \geq 3,$$

which results quickly in overflow. The analysis of the instability this time proceeds as follows. Let

(2.82) $$y_i = \lambda^i, \ \lambda \text{ a constant}, \ i = 1, 2, \ldots,$$

be substituted into (2.77), so that

$$2\lambda^{i+2} - 5\lambda^{i+1} + 2\lambda^i = 0.$$

For $\lambda \neq 0$, the latter equation implies

INSTABILITY

$$2\lambda^2 - 5\lambda + 2 = 0$$

the roots of which are

$$\lambda_1 = 2, \lambda_2 = \frac{1}{2}.$$

In analogy with the construction of the general solution of a second order linear differential equation, one has that

(2.83) $$y_i = c_1(2)^i + c_2(\frac{1}{2})^i$$

is a solution of (2.77) for any constants c_1 and c_2. Now, (2.78) implies that $c_1 = 0$, $c_2 = \frac{1}{2}$, so that (2.83) reduces to

(2.84) $$y_i = (\frac{1}{2})^{i+1}, \quad i = 0,1,2,\ldots .$$

Formula (2.84) is the well-behaved solution generated first by using exact calculations. But, when calculating with the rounding procedure described above, so that the initial conditions are (2.80)-(2.81), (2.83) yields

(2.85) $$y_i = \frac{1}{30}(2)^i + \frac{14}{30}(\frac{1}{2})^i .$$

The term $\frac{1}{30}(2)^i$ in (2.85) soon dominates the term $\frac{14}{30}(\frac{1}{2})^i$ and results in an overflow which is due strictly to roundoff error.

Thus, we have illustrated two unstable calculations, one of which can be corrected by decreasing the grid size, the other of which cannot be so corrected. For nonlinear problems instability analysis often is not possible, and the method of decreasing the grid size is usually the first step one takes in trying to eliminate instability which is not due to programming or machine error.

The problem of instability is fundamental in the study of any initial value problem. Hence, it has received extensive attention and the interested reader should examine such topics as pointwise, local, strong, weak, and stepwise instability, and the intimate relationships between convergence and stability in such texts as Hildebrand (2).

2.7 Periodic Solutions of van der Pol's Equation

Though periodic solutions of nonlinear differential equations have long been of interest in applied science, the application of classical mathematical techniques has limited inquiry largely to questions relating, for example, to existence and uniqueness of such solutions (see, e.g., Cesari). Let us show then how to apply numerical methods to approximate periodic solutions of a classical oscillation equation, the van der Pol equation,

(2.86) $$\ddot{x} - \lambda(1 - x^2)\dot{x} + x = 0; \quad \lambda > 0,$$

where differentiation is with respect to t and where λ is a constant. The choice of the variables x and t in (2.86) is for consistency with the extensive literature on this equation.

Since it is known that (2.86) has a unique periodic solution for each λ, we consider the problem as that of determining the constant α for which the solution of the initial value problem defined by (2.86) and

(2.87) $$x(0) = \alpha, \quad \dot{x}(0) = 0$$

is periodic. For this purpose, let $T = T(\lambda)$ represent the period of the solution which is sought and note (Clenshaw) that

(2.88) $$x(T/2) = -\alpha, \quad \dot{x}(T/2) = 0 .$$

Since Runge-Kutta formulas were applied in Section 2.5, for variety let us show how to apply an eighth-order Taylor series method to the present problem. For this purpose set

(2.89) $$x_{k+1} = x_k + \frac{h}{1}\dot{x}_k + \frac{h^2}{2!}\ddot{x}_k + \frac{h^3}{3!}\dddot{x}_k + \frac{h^4}{4!}x_k^{iv} + \frac{h^5}{5!}x_k^{v} + \frac{h^6}{6!}x_k^{vi} + \frac{h^7}{7!}x_k^{vii} + \frac{h^8}{8!}x_k^{viii} ,$$

(2.90) $$\dot{x}_{k+1} = \dot{x}_k + \frac{h}{1}\ddot{x}_k + \frac{h^2}{2!}\dddot{x}_k + \frac{h^3}{3!}x_k^{iv} + \frac{h^4}{4!}x_k^{v} + \frac{h^5}{5!}x_k^{vi} + \frac{h^6}{6!}x_k^{vii} + \frac{h^7}{7!}x_k^{viii} + \frac{h^8}{8!}x_k^{ix} ,$$

where, from (2.86),

$$\dddot{x}_k = \lambda \ddot{x}_k - \lambda x_k^2 \dot{x}_k - x_k$$

$$\ddddot{x}_k = \lambda \dddot{x}_k - 2\lambda x_k (\dot{x}_k)^2 - \lambda x_k^2 \ddot{x}_k - \dot{x}_k$$

$$x_k^{iv} = \lambda \ddddot{x}_k - 2\lambda (\dot{x}_k)^3 - 6\lambda x_k \dot{x}_k \ddot{x}_k - \lambda x_k^2 \dddot{x}_k - \ddot{x}_k$$

$$x_k^v = \lambda x_k^{iv} - 12\lambda (\dot{x}_k)^2 \ddot{x}_k - 6\lambda x_k (\ddot{x}_k)^2 - 8\lambda x_k \dot{x}_k \dddot{x}_k - \lambda x_k^2 x_k^{iv} - \dddot{x}_k$$

$$x_k^{vi} = \lambda x_k^v - 30\lambda \dot{x}_k (\ddot{x}_k)^2 - 20\lambda (\dot{x}_k)^2 \dddot{x}_k - 20\lambda x_k \ddot{x}_k \dddot{x}_k - 10\lambda x_k \dot{x}_k x_k^{iv}$$
$$- \lambda x_k^2 x_k^v - x_k^{iv}$$

$$x_k^{vii} = \lambda x_k^{vi} - 30\lambda (\ddot{x}_k)^3 - 120\lambda \dot{x}_k \ddot{x}_k \dddot{x}_k - 30\lambda (\dot{x}_k)^2 x_k^{iv} - 20\lambda x_k (\dddot{x}_k)^2$$
$$- 30\lambda x_k \ddot{x}_k x_k^{iv} - 12\lambda x_k \dot{x}_k x_k^v - \lambda x_k^2 x_k^{vi} - x_k^v$$

$$x_k^{viii} = \lambda x_k^{vii} - 210\lambda (\ddot{x}_k)^2 \dddot{x}_k - 140\lambda \dot{x}_k (\dddot{x}_k)^2 - 210\lambda \dot{x}_k \ddot{x}_k x_k^{iv}$$
$$- 42\lambda (\dot{x}_k)^2 x_k^v - 42\lambda x_k \ddot{x}_k x_k^v - 70\lambda x_k \dddot{x}_k x_k^{iv} - 14\lambda x_k \dot{x}_k x_k^{vi}$$
$$- \lambda x_k^2 x_k^{vii} - x_k^{vi}$$

$$x_k^{ix} = \lambda x_k^{viii} - 560\lambda \ddot{x}_k (\dddot{x}_k)^2 - 420\lambda (\ddot{x}_k)^2 x_k^{iv} - 560\lambda \dot{x}_k \dddot{x}_k x_k^{iv}$$
$$- 336\lambda \dot{x}_k \ddot{x}_k x_k^v - 56\lambda (\dot{x}_k)^2 x_k^{vi} - 112\lambda x_k \dddot{x}_k x_k^v - 56\lambda x_k \ddot{x}_k x_k^{vi}$$
$$- 70\lambda x_k (x_k^{iv})^2 - 16\lambda x_k \dot{x}_k x_k^{vii} - \lambda x_k^2 x_k^{viii} - x_k^{vii} .$$

The method proceeds as follows. Let $x_0^{(n)} = n + 1$, $n = 0, 1, 2, \ldots, 10$. For each such $x_0^{(n)}$ generate, in order, sequence $x_{k+1}^{(n)}$, $k = 0, 1, 2, \ldots$ from (2.87), (2.89), (2.90). Initially, each of these sequences will be a decreasing sequence and will decrease from the given positive value down through negative values. Terminate each iteration when

$$x_{K+1}^{(n)} \geq x_K^{(n)},$$

that is, when each sequence stops decreasing, and record

$$S_n = x_0^{(n)} + x_K^{(n)}.$$

Of course, K depends on n. The finite sequence S_n, n = 0,1,2, ...,10, will be an increasing sequence which, initially, is negative. From the first condition of (2.88), we seek to find a negative $x_K^{(n)}$ so that S_n is zero. With this in mind, let n = μ be the first value of n for which

$$S_\mu \cdot S_{\mu+1} \leq 0.$$

Then set $\alpha = x_0^{(\mu)}$ and $T/2 = K\Delta t$. Thus, $x_0^{(\mu)}$ is an integer which approximates α. To compute a one decimal place refinement of this approximation, set $x_0^{(0)} = 0.0 + x_0^{(\mu)}$, $x_0^{(1)} = 0.1 + x_0^{(\mu)}$, $x_0^{(2)} = 0.2 + x_0^{(\mu)}$, ..., $x_0^{(10)} = 1.0 + x_0^{(\mu)}$, and recycle. Thus, if one had found $x_0^{(\mu)} = 2$, one would recycle with $x_0^{(0)} = 2.0$, $x_0^{(1)} = 2.1$, $x_0^{(2)} = 2.2,\ldots,x_0^{(10)} = 3.0$. From the resulting one decimal place refinement, one can, in the indicated fashion, construct a two decimal place refinement, and, in the same manner, a j-decimal place refinement, where the magnitude of j is limited only by one's computer capability.

56   ORDINARY DIFFERENTIAL EQUATIONS

On the UNIVAC 1108, the following approximations for α and $T/2$ were generated by the method of this section with $\Delta t = 0.001$:

$$\lambda = 0.1 \quad \alpha = 2.000 \quad T/2 = 3.148$$
$$\lambda = 1.0 \quad \alpha = 2.009 \quad T/2 = 3.335$$
$$\lambda = 10 \quad \alpha = 2.014 \quad T/2 = 9.538.$$

The graphs of the approximate periodic functions are shown in Figure 2.3. All computations were done in double precision and the total computing time was under 30 minutes.

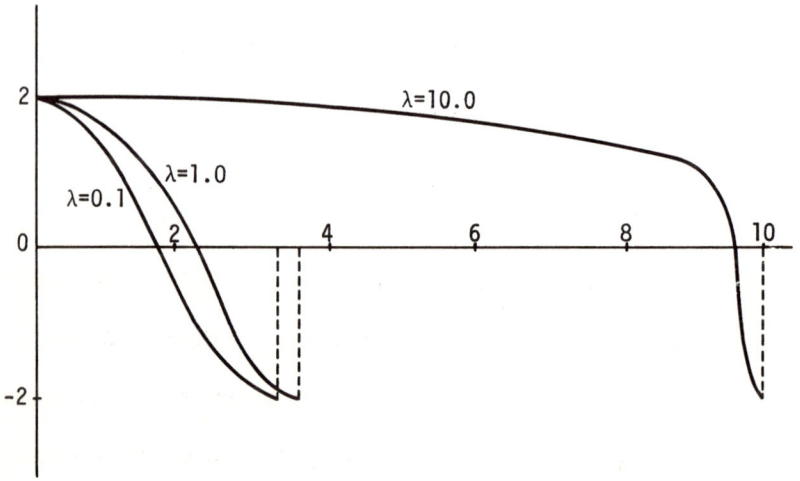

Figure 2.3

2.8 Approximate Solution of Boundary Value Problems

Unlike the solution of an initial value problem, the solution of a boundary value problem, when it exists, need not be unique. Thus, for example, any function of the form $y = c \sin x$, where c is a constant, is a solution of the boundary value problem defined by $y'' + y = 0$, $y(0) = y(\pi) = 0$. Hence, in studying the approximate solution of boundary value problems, let us consider first, for simplicity, the linear problem defined by

(2.91) $$y'' + P(x)y' + Q(x)y = R(x)$$

(2.92) $$y(a) = \alpha, \quad y(b) = \beta, \quad a < b,$$

where $P(x)$, $Q(x)$, $R(x)$ are continuous on $a \leq x \leq b$, and let us assume, in addition, that

(2.93) $$Q(x) \leq 0, \quad a < x < b,$$

which is sufficient to imply that the solution of (2.91)-(2.92) exists and is unique (see, e.g., Keller).

Problem (2.91)-(2.93) can be discretized by replacing the interval $a \leq x \leq b$, which consists of an infinite number of points, by a finite $G[a,h]$ set whose terminal point is $x = b$, and by replacing the second order differential equation (2.91) by a second order difference equation. To do this, divide $a \leq x \leq b$ into n equal parts, each of length $\Delta x = (b-a)/n$ by the set $a = x_0 < x_1 < x_2 \cdots < x_n = b$, where $x_j = a + j\Delta x$, $j = 0,1,2,\ldots,n$, and substitute (2.9) and any one of (2.4)-(2.8) into (2.91) to obtain an approximating difference equation at $x_1, x_2, x_3, \ldots, x_{n-1}$. Thus, using (2.6), which is popular currently, one would have

$$\frac{y_{i-1} - 2y_i + y_{i+1}}{(\Delta x)^2} + P(x_i) \frac{y_{i+1} - y_{i-1}}{2\Delta x} + Q(x_i)y_i = R(x_i), \quad i = 1, 2, \ldots, n-1,$$

or, equivalently,

(2.94) $\quad y_{i-1}[2 - P(x_i)\Delta x] + y_i[-4 + 2(\Delta x)^2 Q(x_i)] + y_{i+1}[2 + P(x_i)\Delta x]$

$$= 2(\Delta x)^2 R(x_i); \quad i = 1, 2, \ldots, n-1 .$$

From (2.92), it follows that if one writes down, <u>in order</u>, each of equations (2.94), as will be illustrated next, there results a tridiagonal linear system of $(n-1)$ equations in the $(n-1)$ unknowns $y_1, y_2, \ldots, y_{n-1}$. A solution of this system constitutes an approximate solution of problem (2.91)-(2.93) at $x_1, x_2, \ldots, x_{n-1}$.

<u>Example 1</u>

Consider the relatively simple boundary value problem

(2.95) $\qquad\qquad y'' - y = 1, \quad 0 < x < 3$

(2.96) $\qquad\qquad y(0) = -1, \; y(3) = -1 .$

BOUNDARY VALUE PROBLEMS

With $\Delta x = \frac{1}{2}$, divide $0 \leq x \leq 3$ into six equal parts so that $x_0 = 0$, $x_1 = 0.5$, $x_2 = 1$, $x_3 = 1.5$, $x_4 = 2$, $x_5 = 2.5$, and $x_6 = 3$. Since no approximation for y' must be made, (2.97) is approximated by the difference equation

(2.97) $$\frac{y_{i-1} - 2y_i + y_{i+1}}{(\Delta x)^2} - y_i = 1, \quad i = 1,2,3,4,5,$$

or, equivalently, by

(2.98) $$y_{i-1} - \frac{9}{4}y_i + y_{i+1} = \frac{1}{4}, \quad i = 1,2,3,4,5.$$

Using the given boundary values, the tridiagonal system that results for $i = 1,2,3,4,5$ is

(2.99)
$$-\frac{9}{4}y_1 + y_2 = \frac{5}{4}$$
$$y_1 - \frac{9}{4}y_2 + y_3 = \frac{1}{4}$$
$$y_2 - \frac{9}{4}y_3 + y_4 = \frac{1}{4}$$
$$y_3 - \frac{9}{4}y_4 + y_5 = \frac{1}{4}$$
$$y_4 - \frac{9}{4}y_5 = \frac{5}{4},$$

the exact solution of which is $y_1 = y_2 = y_3 = y_4 = y_5 = -1$. It is not surprising, because of the simplicity of the problem, that the numerical solution happens to coincide with the exact solution, $y = -1$, at the grid points. It is <u>most important</u> to note, however, with regard to the matrix of system (2.99), that the coefficient of y_i in (2.98) yields the main diagonal entries, that of y_{i-1} yields the entries just below the main diagonal, and that of y_{i+1} yields the entries just above the main diagonal. These observations also extend to (2.94).

Example 2

Consider the general linear boundary value problem (2.91)-(2.93) and let (2.91) be approximated by (2.94). If $a \leq x \leq b$ is divided into n equal parts, as in Example 1, it follows that (2.94) yields a tridiagonal linear system. It is not at all obvious, however, that this system has a unique solution. Let us see if the system can be made diagonally dominant so that we can have this assurance. From (2.94), we wish to have

$$(2.100) \qquad |-4 + 2(\Delta x)^2 Q(x_i)| \geq |2 + \Delta x\, P(x_i)| + |2 - \Delta x\, P(x_i)|.$$

For those values of Δx which satisfy both

$$(2.101) \qquad 2 - (\Delta x) P(x_i) > 0, \quad 2 + (\Delta x) P(x_i) > 0,$$

condition (2.100) reduces to

$$4 - 2(\Delta x)^2 Q(x_i) \geq (2 + (\Delta x) P(x_i)) + (2 - (\Delta x) P(x_i)),$$

or

BOUNDARY VALUE PROBLEMS

(2.102) $$-2(\Delta x)^2 Q(x_i) \geq 0.$$

Since $Q(x_i) \leq 0$, (2.102) is valid. Thus, it is sufficient to satisfy (2.101), or, equivalently,

$$\Delta x \, |P(x_i)| < 2.$$

If M is any upper bound for $|P(x)|$ on $a \leq x \leq b$, so that

$$|P(x)| \leq M, \quad a \leq x \leq b,$$

then a sufficient condition to assure diagonal dominance is, finally,

(2.103) $$\Delta x < 2/M.$$

Example 3

Consider the general linear boundary value problem (2.91)-(2.93) again. In Example 2 it was shown that if one approximates (2.91) by (2.94) and if one wishes to have diagonal dominance, then the number of parts into which one divides $a \leq x \leq b$ is no longer arbitrary. Let us show now how (2.91)-(2.93) can be approximated somewhat less accurately, but in a fashion that one is free to divide the interval into an arbitrary number of parts and still obtain diagonal dominance. Suppose that (2.91) is approximated at each point x_i, $i = 1, 2, \ldots, n-1$, as follows. Approximate y" by (2.9). To approximate y', first examine $P(x_i)$. If $P(x_i) \geq 0$, use forward difference approximation (2.4), while if $P(x_i) < 0$, use backward difference approximation (2.5).

Thus, if $P(x_i) \geq 0$, there results

$$\frac{y_{i-1} - 2y_i + y_{i+1}}{(\Delta x)^2} + P(x_i) \frac{y_{i+1} - y_i}{\Delta x} + Q(x_i)y_i = R(x_i),$$

or, equivalently,

$$y_{i-1} + y_i[-2 - P(x_i)\Delta x + Q(x_i)(\Delta x)^2] + y_{i+1}[1 + P(x_i)\Delta x] = (\Delta x)^2 R(x_i),$$

while if $P(x_i) < 0$, there results

$$\frac{y_{i-1} - 2y_i + y_{i+1}}{(\Delta x)^2} + P(x_i) \frac{y_i - y_{i-1}}{\Delta x} + Q(x_i)y_i = R(x_i),$$

or, equivalently,

$$y_{i-1}[1 - P(x_i)\Delta x] + y_i[-2 + P(x_i)\Delta x + Q(x_i)(\Delta x)^2] + y_{i+1} = (\Delta x)^2 R(x_i).$$

Such an approach <u>always</u> yields diagonal dominance of the resulting system, for it is the coefficient of y_i which determines the main diagonal elements of the resulting linear system. The scheme presented in this example is called a <u>forward-backward</u> technique.

With regard to Examples 1-3, above, note that SOR is known to converge for all initial guesses and for all ω in the range $0 < \omega < 2$ when the system which results is diagonally dominant (see, e.g., Varga). Though results for determining the optimum ω are known, these are usually nonconstructive, so that in practice one usually selects a set of ω's in the range $0 < \omega < 2$, lets each

BOUNDARY VALUE PROBLEMS

run for, say, fifteen iterations with a zero initial vector, and then chooses that ω which seems to be giving the most rapid convergence.

To extend the ideas presented thus far to nonlinear problems, consider the general nonlinear boundary value problem

(2.104) $$y'' = f(x,y,y')$$

(2.105) $$y(a) = \alpha, \quad y(b) = \beta; \quad a < b,$$

where a, b, α, β are constants and $f(x,y,z)$ has continuous first order partial derivatives for all y, for all z, and for x in the range $a \leq x \leq b$. For convenience, we will assume that

(2.106) $$\frac{\partial f(x,y,z)}{\partial y} \geq 0, \quad \left|\frac{\partial f(x,y,z)}{\partial z}\right| \leq M,$$

where M is a positive constant.

Boundary value problem (2.104)-(2.106) has a unique solution (see, e.g., Keller). Moreover, when (2.104) has the linear form (2.91), the first condition of (2.106) is exactly (2.93).

Problem (2.104)-(2.106) is descretized by replacing the interval $a \leq x \leq b$ by a finite $G[a,h]$ set, as in the linear case, and by approximating differential equation (2.104) by a difference equation which uses (2.6) and (2.9), that is, by

(2.107) $$\frac{y_{i-1} - 2y_i + y_{i+1}}{(\Delta x)^2} = f(x_i, y_i, \frac{y_{i+1} - y_{i-1}}{2\Delta x}), \quad i = 1, 2, \ldots, n-1.$$

One writes down, in order, each of (2.107) for $i = 1, 2, \ldots, n-1$, which, by (2.105), results in a system of nonlinear algebraic or transcendental equations in $y_1, y_2, \ldots, y_{n-1}$. A solution of this system constitutes an approximate solution of (2.104)-(2.106) on the grid points $x_1, x_2, \ldots, x_{n-1}$.

Example

For the one dimensional radiation equation

$$(2.108) \qquad y'' - e^y = 0 ,$$

consider the boundary value problem with $y(0) = y(1) = 0$. The problem has a unique solution which can be approximated as follows. Divide $0 \leq x \leq 1$ into three equal parts by the points $x_0 = 0$, $x_1 = \frac{1}{3}$, $x_2 = \frac{2}{3}$, $x_3 = 1$. Approximate (2.108) by

$$\frac{y_{i-1} - 2y_i + y_{i+1}}{(\frac{1}{3})^2} - e^{y_i} = 0; \quad i = 1, 2 .$$

Thus,

$$9y_0 - 18y_1 + 9y_2 - e^{y_1} = 0$$

$$9y_1 - 18y_2 + 9y_3 - e^{y_2} = 0 ,$$

so that, inserting the boundary conditions, one has

$$e^{y_1} + 18y_1 - 9y_2 = 0$$

$$e^{y_2} + 18y_2 - 9y_1 = 0.$$

This system is similar to that solved by (1.37) – (1.38).

The iteration takes the particular form

$$y_1^{(k+1)} = y_1^{(k)} - \omega [e^{y_1^{(k)}} + 18y_1^{(k)} - 9y_2^{(k)}]/[e^{y_1^{(k)}} + 18]$$

$$y_2^{(k+1)} = y_2^{(k)} - \omega [e^{y_2^{(k)}} + 18y_2^{(k)} - 9y_1^{(k+1)}]/[e^{y_2^{(k)}} + 18],$$

and can be executed easily on a high speed digital computer.

Note that the nonlinear system generated by the numerical method of this section can be proved to have a unique solution only for all sufficiently small Δx (Bers) and that the resulting equations can always be solved by the generalized Newton's method for all ω in some proper subset of the range $0 < \omega < 2$ (S. Schechter). For a variety of sufficient conditions which assure that numerical solutions of both linear and nonlinear problems converge to the analytical solutions as $\Delta x \to 0$, see Keller.

2.9 Remarks

There exist a variety of other discrete methods which can be of value in the numerical solution of initial value problems. General categories include numerical integration, predictor-corrector, and multistep methods. There also exist valuable special methods for special problems, like those relating to stiff equations and to second order equations which do not contain y'. Finally, note that often one can conveniently solve a boundary value problem by an initial value technique and an initial value problem by a boundary value technique. (General reading sources for the above material are Collatz (1), Fox (1,2), Greenspan (4,6), Hamming, Henrici, Noble, Ralston, and Todd).

Exercises

1. Develop error estimates for approximations (2.4)-(2.9).

2. Find y_1, y_2, and y_3 by the method of Taylor series for each of the following initial value problems. Use a third order series approximation with $h = 0.5$. Whenever possible, find the exact solution of the problem and compare it with the numerical solution.

   (a) $y'' + y = 0$, $y(0) = 0$, $y'(0) = 1$

   (b) $y'' - 3y' + 2y = 0$, $y(0) = 1$, $y'(0) = 0$

   (c) $y'' - 4y' = x^2 - 2$, $y(0) = 1$, $y'(0) = 1$

   (d) $y'' = y^2 + x^2$, $y(0) = 1$, $y'(0) = 0$

   (e) $y'' = e^y$, $y(0) = 1$, $y'(0) = 0$

   (f) $y'' + (y')^2 = x$, $y(-1) = 0$, $y'(-1) = 1$

   (g) $y'' + 3y' - \cos y = 0$, $y(0) = 1$, $y'(0) = -1$

   (h) $y'' + y^3 = e^y$, $y(0) = 1$, $y'(0) = 2$.

3. Show that Heun's formula approximates (2.35) through terms of order h^3.

4. Find y_1, y_2, and y_3 for each initial value problem of Exercise 2 using Kutta's formulas (2.56).

5. Consider the following initial value problems:

   (a) $y' = -y$, $y(0) = 1$

   (b) $y' = -100y$, $y(0) = 1$

   (c) $y' = -1000y$, $y(0) = 10$.

For what values of Δx will Euler's method be unstable?

6. Approximate the periodic solution of van der Pol's equation for each of $\lambda = 0.001$, $\lambda = 5.0$, $\lambda = 20.0$.

7. Find an approximate solution with $h = 0.2$ for each of the following boundary value problems. Whenever possible, find the exact solution of the problem and compare the numerical solution with it.

(a) $y'' - y = 0$, $y(0) = 1$, $y(\frac{\pi}{6}) = 0$

(b) $y'' + 3y' - 4y = 0$, $y(0) = 0$, $y(1) = 1$

(c) $y'' - 5y' + 4y = x^2 - 2x + 1$, $y(0) = 1$, $y(1) = -1$

(d) $y'' - 3xy' - y = x^2$, $y(0) = 1$, $y(50) = -1$

(e) $y'' - (25 - x)y' - y = x^2$, $y(0) = 1$, $y(20) = -1$

(f) $y'' - 4xy' + (4x^2 - 2)y = 0$, $y(0) = 1$, $y(1) = 0$

(g) $y'' = y^3 + x$, $y(1) = 1$, $y(2) = -1$

(h) $y'' - 3xy' - y^3 = 0$, $y(1) = 1$, $y(2) = 0$

(i) $y'' = e^y$, $y(0) = 1$, $y(1) = 0$

(j) $y'' - 4xy' = e^y$, $y(0) = 1$, $y(2) = -1$.

CHAPTER III

NUMERICAL SOLUTION OF ELLIPTIC BOUNDARY VALUE PROBLEMS

3.1 Introduction

The natural generalization of a second order <u>ordinary</u> differential equation is a second order <u>partial</u> differential equation, which can be defined in two dimensions as follows. On a plane point set R, the equation

$$(3.1) \quad a(x,y,u,\frac{\partial u}{\partial x},\frac{\partial u}{\partial y})\frac{\partial^2 u}{\partial x^2} + 2b(x,y,u,\frac{\partial u}{\partial x},\frac{\partial u}{\partial y})\frac{\partial^2 u}{\partial x \partial y}$$

$$+ c(x,y,u,\frac{\partial u}{\partial x},\frac{\partial u}{\partial y})\frac{\partial^2 u}{\partial y^2} + f(x,y,u,\frac{\partial u}{\partial x},\frac{\partial u}{\partial y}) = 0,$$

subject to the restriction that at each point of R

$$(3.2) \quad a^2 + b^2 + c^2 \neq 0,$$

is called a second order, quasilinear partial differential equation. In the special case when

$$a = a(x,y), \quad b = b(x,y), \quad c = c(x,y),$$

then (3.1) is called <u>linear</u> if

$$f \equiv d(x,y)\frac{\partial u}{\partial x} + e(x,y)\frac{\partial u}{\partial y} + g(x,y)u + h(x,y),$$

while it is called <u>mildly nonlinear</u> if

$$f \equiv f(x,y,u).$$

We will consider no second order equation more complex than (3.1).

It is convenient for both practical and theoretical reasons to categorize various second order partial differential equations as follows. At a given point of definition in the plane, equation (3.1), in analogy with the conic sections, is said to be

elliptic if $b^2 - ac < 0$,

parabolic if $b^2 - ac = 0$,

hyperbolic if $b^2 - ac > 0$,

Example 1

At each point in the plane, the equation

$$(3.3) \qquad \frac{\partial^2 u}{\partial x^2} + \frac{\partial^2 u}{\partial y^2} = 0$$

is elliptic. This equation is called the <u>potential</u> equation, or Laplace's equation, and is the prototype elliptic partial differential equation.

Example 2

At each point in the plane, the equation

$$(3.4) \qquad \frac{\partial^2 u}{\partial x^2} - \frac{\partial u}{\partial y} = 0$$

INTRODUCTION

is parabolic. This equation is called the *heat* equation and is the prototype parabolic partial differential equation.

Example 3

At each point in the plane, the equation

$$\text{(3.5)} \qquad \frac{\partial^2 u}{\partial x^2} - \frac{\partial^2 u}{\partial y^2} = 0$$

is hyperbolic. This equation is called the *wave* equation and is the prototype hyperbolic partial differential equation.

Example 4

At each point in the plane, the minimal surface, or soap film, equation

$$\text{(3.6)} \qquad [1 + (\frac{\partial u}{\partial y})^2]\frac{\partial^2 u}{\partial x^2} - 2\frac{\partial u}{\partial x}\frac{\partial u}{\partial y}\frac{\partial^2 u}{\partial x \partial y} + [1 + (\frac{\partial u}{\partial x})^2]\frac{\partial^2 u}{\partial y^2} = 0$$

is elliptic, since

$$b^2 - ac = (\frac{\partial u}{\partial x})^2 (\frac{\partial u}{\partial y})^2 - [1 + (\frac{\partial u}{\partial x})^2][1 + (\frac{\partial u}{\partial y})^2]$$

$$= -1 - (\frac{\partial u}{\partial x})^2 - (\frac{\partial u}{\partial y})^2 < 0.$$

Example 5

At each point in the plane consider the gas dynamical equation

$$(3.7) \quad [c_0^2 - (\frac{\partial u}{\partial x})^2]\frac{\partial^2 u}{\partial x^2} - 2\frac{\partial u}{\partial x}\frac{\partial u}{\partial y}\frac{\partial^2 u}{\partial x \partial y} + [c_0^2 - (\frac{\partial u}{\partial y})^2]\frac{\partial^2 u}{\partial y^2} = 0,$$

where c_0 is the speed of sound. Then

$$b^2 - ac = (\frac{\partial u}{\partial x})^2(\frac{\partial u}{\partial y})^2 - [c_0^2 - (\frac{\partial u}{\partial x})^2][c_0^2 - (\frac{\partial u}{\partial y})^2]$$

$$= -c_0^4 \{1 - \frac{1}{c_0^2}[(\frac{\partial u}{\partial x})^2 + (\frac{\partial u}{\partial y})^2]\}.$$

Let the nonnegative number M, called the Mach number, be defined by

$$M = \{\frac{1}{c_0^2}[(\frac{\partial u}{\partial x})^2 + (\frac{\partial u}{\partial y})^2]\}^{1/2}.$$

Then,

$$b^2 - ac = c_0^4[M^2 - 1].$$

Thus, if $M < 1$, equation (3.7) is elliptic and the corresponding flow is called subsonic; if $M = 1$, equation (3.7) is parabolic and the corresponding flow is called sonic; and, if $M > 1$, equation (3.7) is hyperbolic and the corresponding flow is called supersonic.

Often it will be of value to use the notation

$$u_x = \frac{\partial u}{\partial x}, \; u_y = \frac{\partial u}{\partial y}, \; u_{xx} = \frac{\partial^2 u}{\partial x^2}, \; u_{xy} = \frac{\partial^2 u}{\partial x \partial y}, \; u_{yy} = \frac{\partial^2 u}{\partial y^2}, \ldots,$$

LAPLACE'S EQUATION

so that, for example, (3.7) can be written in the following, more compact form:

$$(c_0^2 - u_x^2)u_{xx} - 2u_x u_y u_{xy} + (c_0^2 - u_y^2)u_{yy} = 0.$$

Note that the character or type of any second order quasilinear partial differential equation is determined <u>completely</u> by the coefficients of its second order terms.

Elliptic equations will be studied in this chapter, parabolic equations in Chapter 4 and hyperbolic equations in Chapter 5. A more general approach to hyperbolic equations in terms of systems will be developed in Chapter 7.

3.2 Boundary Value Problems for the Laplace Equation

Let us begin the study of elliptic equations by considering the simplest such equation, that is, Laplace's equation (3.3). Alternate ways of writing this equation are

(3.8) $$\begin{cases} u_{xx} + u_{yy} = 0 \\ \Delta u = 0 \\ \nabla^2 u = 0. \end{cases}$$

Any solution of Laplace's equation is called a <u>harmonic</u> function and the special properties of harmonic functions, because of their importance in the study of gravitation and potential theory, have been studied in great detail. One such property, for example, is the <u>max-min property</u>, which can be stated as follows: if R is a bounded, simply connected region whose boundary is S, and if u is harmonic on R and continuous on R + S, then u takes on its maximum and its minimum values on S. That the max-min property for harmonic functions is reasonable can be seen from the following example. Consider the particular function $u = x^2 - y^2$, which is everywhere continuous and harmonic, and whose graph is the saddle surface shown in Figure 3.1. Let Q be the point on the surface whose coordinates are (1,0,1). Then the intersection of the surface with the plane whose equation is $y = 0$ is a parabola P_1, which opens upward, while the intersection of the surface with the plane whose equation is $x = 1$ is a parabola P_2, which opens downward (see Figure 3.1). This opposition of concavities is common to all nonconstant harmonic functions because (3.3) implies that $u_{xx} = -u_{yy}$ at each point of definition. But, point Q in Figure 3.1 cannot be either a maximum or a minimum point, since in any neighborhood of it there are points on the surface which are relatively higher and other points which are relatively lower. Thus, if $u = x^2 - y^2$ is defined only on a bounded, simply connected point set R and on its boundary S, then no maximum or minimum value of u can occur on R, from which the max-min property follows immediately.

LAPLACE'S EQUATION

For other harmonic functions, like $u = 10$, the maximum and minimum values <u>are</u> attained on R, but they are <u>also</u> attained on S. Thus, the max-min property, in general, does not preclude a harmonic function attaining extreme values on both R and S.

The most meaningful types of problems for elliptic equations, from both the physical and the mathematical points of view, are boundary value problems, and the simplest such problem is the Dirichlet problem, which is formulated for the Laplace equation as follows.

<u>Dirichlet Problem</u>

Let G be a bounded point set whose interior R is simply connected and whose boundary S is piecewise regular, that is, is piecewise continuously differentiable. If $f(x,y)$ is given and continuous on S, then the Dirichlet problem for the Laplace equation is that of finding a function $u(x,y)$ which is

(a)   defined and continuous on R + S,

(b)   identical with $f(x,y)$ on S, and

(c)   harmonic on R.

<u>Example</u>

The problem of determining a function $u(x,y)$ such that:

(a)   u is continuous at each point (x,y) whose coordinates satisfy $x^2 + y^2 \leq 25$;

(b)   u coincides with $f(x,y) = \frac{x-y^2}{9}$ at each point (x,y) whose coordinates satisfy $x^2 + y^2 = 25$; and

(c)   u is harmonic at each point (x,y) whose coordinates satisfy $x^2 + y^2 < 25$,

is a Dirichlet problem.

Geometrically, the Dirichlet problem can be interpreted as follows. Since $f(x,y)$ is defined only on S and is continuous on S, the graph of $f(x,y)$ is a closed space curve (see Figure 3.2). In the Dirichlet problem, one is asked to find a function $u(x,y)$ which is harmonic on R and whose graph, which is a surface over R + S, contains the space curve f and has this curve for its boundary.

That the solution of the Dirichlet problem exists and is unique has been proved by a variety of methods, including subharmonic and superharmonic functions, Green's functions, finite differences, Dirichlet's principle, integral equations and conformal mapping (see, e.g., Courant, Friedrichs and Lewy; Courant and Hilbert; Garabedian; Greenspan (2); Kellogg; Petrovsky). The analytical determination of $u(x,y)$, however, is a far more difficult problem than that of establishing its existence and uniqueness. If S is a rectangle, then the solution can be constructed as a Fourier series (Churchill, Greenspan (2), Petrovsky), while if S is a circle or an ellipse, then

LAPLACE'S EQUATION

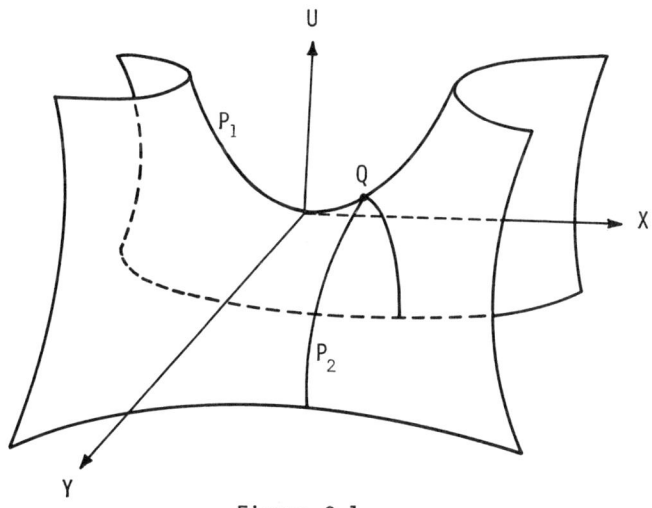

Figure 3.1

Figure 3.2

the solution can be constructed as a Poisson integral or a Fourier series (Churchill, Greenspan (2), Petrovsky, Royster). Also, any problem for which an explicit conformal map can be given which takes R onto a rectangular, circular, or elliptic region can be solved in closed form (Nehari). Beyond these cases, the problems involved in constructing u do not seem to be amenable to existing general analytical techniques.

But, unfortunately, even in the few cases where a solution can be produced as a Fourier series or as an integral, one may not be able to evaluate the solution at a point of interest because the series may be slowly convergent, or because an integrand may not have an antiderivative.

Because it is known that the Dirichlet problem always has a unique solution, because in most problems the solutions cannot be given in closed form, and because in those problems which have closed form solutions the solutions can rarely be evaluated at a particular point of interest, we will seek to develop an algorithm for the approximate solution of the Dirichlet problem. For the present, we conclude this section by discussing other kinds of boundary value problems which are of interest with regard to elliptic equations.

In the statement of the Dirichlet problem, if one were to replace the boundary values $u = f(x,y)$ on S by normal derivative values

LAPLACE'S EQUATION

$\frac{\partial u}{\partial n} = g(x,y)$, then the resulting problem is called a <u>Neumann</u> problem, and such problems, in general, have an infinite number of solutions, any two of which differ only by an additive constant (Courant and Hilbert). In the statement of the Dirichlet problem, if one prescribes function values $u = f(x,y)$ on a nonempty, proper subset of S and normal derivatives $\frac{\partial u}{\partial n} = g(x,y)$ on the remainder of S, then the resulting problem is called a <u>Mixed Type</u> or <u>Robin</u> problem, which, in general, has a unique solution (Courant and Hilbert). A final type problem which is intimately associated with the propagation of waves is the <u>exterior Dirichlet problem</u>, which is formulated precisely as follows.

Exterior Dirichlet Problem

Let G be a bounded point set whose interior R is simply connected and whose boundary S is piecewise regular. Let R* be the exterior of G. If $f(x,y)$ is given and continuous on S, then the <u>exterior Dirichlet problem</u> for the Laplace equation is that of determining a function $u(x,y)$ on R* + S which is

(a) defined and continuous on R* + S

(b) harmonic on R*

(c) identical with $f(x,y)$ on S, and

(d) bounded on R* + S.

It is known that the exterior Dirichlet problem has a unique solution (Courant and Hilbert), but no general analytical technique is available at present for constructing the solution.

3.3 Difference Equation Approximation of Laplace's Equation

In developing a numerical method for the Dirichlet problem, it will be important to have a difference equation approximation of the Laplace equation, and this will be developed first. For $h > 0$ and for

(3.9) $$0 < h_i \leq h; \quad i = 1,2,3,4,$$

let the points (x,y), $(x+h_1,y)$, $(x,y+h_2)$, $(x-h_3,y)$, $(x,y-h_4)$ be numbered 0, 1, 2, 3, 4, respectively, as shown in Figure 3.3. At a point numbered i, denote u by u_i, and let us try to determine parameters $\alpha_0, \alpha_1, \alpha_2, \alpha_3, \alpha_4$ such that at (x,y)

(3.10) $$u_{xx} + u_{yy} \equiv \alpha_0 u_0 + \alpha_1 u_1 + \alpha_2 u_2 + \alpha_3 u_3 + \alpha_4 u_4.$$

Since there are five parameters α_i, one would, in general, seek five independent relationships from which they can be determined. Substitution into (3.10) of Taylor expansions about (x,y) for u_1, u_2, u_3, u_4 and regrouping of terms implies

LAPLACE DIFFERENCE APPROXIMATION

$$(3.11) \quad u_{xx} + u_{yy} \equiv u_0(\alpha_0 + \alpha_1 + \alpha_2 + \alpha_3 + \alpha_4) + u_x(h_1\alpha_1 - h_3\alpha_3)$$

$$+ u_y(h_2\alpha_2 - h_4\alpha_4) + \frac{1}{2} u_{xx}(h_1^2\alpha_1 + h_3^2\alpha_3)$$

$$+ \frac{1}{2} u_{yy}(h_2^2\alpha_2 + h_4^2\alpha_4) + \sum_{1}^{4} [O(\alpha_i h_i^3)].$$

Setting corresponding coefficients of (3.11) equal, one finds

$$(3.12) \quad \begin{cases} \alpha_0 + \alpha_1 + \alpha_2 + \alpha_3 + \alpha_4 = 0 \\ h_1\alpha_1 \quad\quad\quad - h_3\alpha_3 \quad\quad = 0 \\ \quad\quad h_2\alpha_2 \quad\quad\quad - h_4\alpha_4 = 0 \\ h_1^2\alpha_1 \quad\quad\quad + h_3^2\alpha_3 \quad\quad = 2 \\ \quad\quad h_2^2\alpha_2 \quad\quad\quad + h_4^2\alpha_4 = 2, \end{cases}$$

the unique solution of which is

$$(3.13) \quad \alpha_0 = -2\left[\frac{1}{h_1 h_3} + \frac{1}{h_2 h_4}\right], \; \alpha_1 = \frac{2}{h_1(h_1 + h_3)}, \; \alpha_2 = \frac{2}{h_2(h_2 + h_4)},$$

$$\alpha_3 = \frac{2}{h_3(h_1 + h_3)}, \; \alpha_4 = \frac{2}{h_4(h_2 + h_4)}.$$

Substitution of (3.13) into (3.10) implies, then, that at (x,y)

$$(3.14) \quad u_{xx} + u_{yy} \equiv -2\left[\frac{1}{h_1 h_3} + \frac{1}{h_2 h_4}\right] u_0 + \frac{2}{h_1(h_1+h_3)} u_1 + \frac{2}{h_2(h_2+h_4)} u_2$$

$$+ \frac{2}{h_3(h_1+h_3)} u_3 + \frac{2}{h_4(h_2+h_4)} u_4 + \sum_{1}^{4} [O(h_i)].$$

Since $\lim_{h \to 0} \sum_{i=1}^{4} [O(h_i)] = 0$, it follows from (3.14) that the approximation

(3.15) $\quad u_{xx} + u_{yy} \sim -2 \left[\dfrac{1}{h_1 h_3} + \dfrac{1}{h_2 h_4} \right] u_0 + \dfrac{2}{h_1(h_1+h_3)} u_1 + \dfrac{2}{h_2(h_2+h_4)} u_2$

$\quad + \dfrac{2}{h_3(h_1+h_3)} u_3 + \dfrac{2}{h_4(h_2+h_4)} u_4$

is reasonable. From (3.15), the difference equation approximation of Laplace's equation which we will use is

(3.16) $\quad -2 \left[\dfrac{1}{h_1 h_3} + \dfrac{1}{h_2 h_4} \right] u_0 + \dfrac{2}{h_1(h_1+h_3)} u_1 + \dfrac{2}{h_2(h_2+h_4)} u_2 + \dfrac{2}{h_3(h_1+h_3)} u_3$

$\quad + \dfrac{2}{h_4(h_2+h_4)} u_4 = 0.$

In the important special case when $h_1 = h_2 = h_3 = h_4 = h$, (3.16) reduces to

(3.16a) $\quad\quad\quad -4u_0 + u_1 + u_2 + u_3 + u_4 = 0.$

Note that the numbering 0, 1, 2, 3, 4 is not essential to the form of (3.16). Thus, if 0, 1, 2, 3, 4 were replaced by 11, 5, 3, 6, 9, respectively, then (3.16) need be altered only by replacing u_0, u_1, u_2, u_3, u_4 with $u_{11}, u_5, u_3, u_6, u_9$, respectively. Note also that (3.16) implies the existence of a discrete max-min property. This can be seen readily from the special form (3.16a), which, re-

LATTICE POINTS

written as

$$u_0 = (u_1 + u_2 + u_3 + u_4)/4,$$

implies that u_0 is the arithmetic mean of u_1, u_2, u_3, u_4, so that

$$\min[u_1, u_2, u_3, u_4] \leq u_0 \leq \max[u_1, u_2, u_3, u_4].$$

Finally, it is important to realize that (3.16) is an algebraic equation which approximates differential equation (3.3), and that the method used in its derivation will apply equally well for differential equations of various types in an arbitrary number of dimensions.

3.4 Interior and Boundary Lattice Points

Consider next discretizing the point set $G = R + S$ given in the statement of the Dirichlet problem. For illustrative purposes,

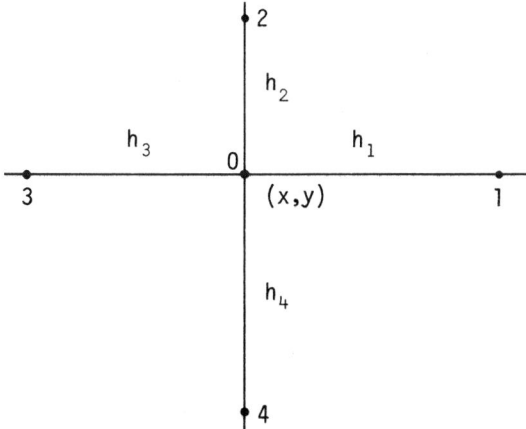

Figure 3.3

if R and S are as shown in Figure 3.4(a), then we wish to replace R by the finite set R_h, which is shown as the set of crossed points in Figure 3.4(b), while we wish to replace S by the finite set S_h, which is shown as the set of circled points in Figure 3.4(b). These finite sets can be defined precisely as follows.

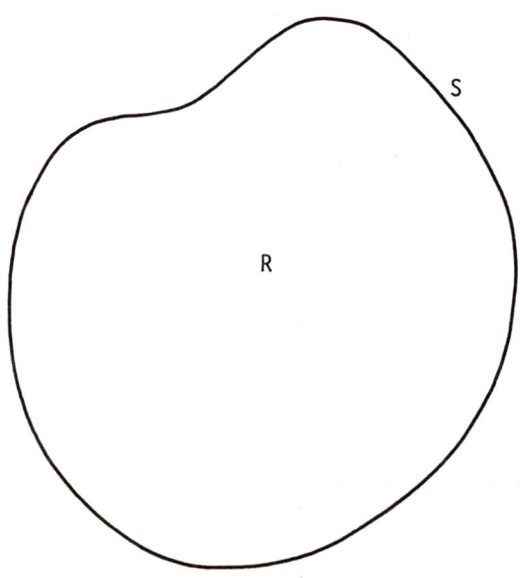

Figure 3.4(a)

LATTICE POINTS 85

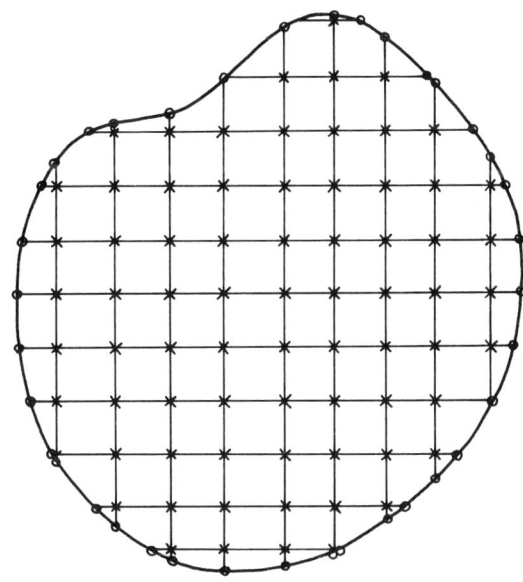

Figure 3.4(b)

Let $(\bar{x},\bar{y})$ be an arbitrary, but fixed, point in the plane, and let h be a positive constant called the grid size. The set of points $(\bar{x}+ph,\bar{y}+qh)$, $p = 0,\pm 1,\pm 2,\ldots, q = 0,\pm 1,\pm 2,\ldots,$ is called a set of planar grid points. The set of vertical lines $x = \bar{x} + ph$ and of horizontal lines $y = \bar{y} + qh$ is called a planar lattice. Those planar grid points which are also points of R are called interior lattice, or grid, points and are denoted by R_h. Let the set of points which S and the planar lattice have in common be denoted by S_h^* and set $G_h^* = R_h + S_h^*$. The four <u>neighbors</u> of a point (x,y) in R_h are defined to be those four points in G_h^* which are closest to (x,y) in the east, north, west, and south directions. Let G_h be that subset of G_h^* which consists of each point of R_h and its four neighbors. Finally, the boundary lattice, or grid, points, denoted by S_h, are defined by $S_h = G_h - R_h$.

Example

Consider the quadrilateral with vertices $(0,0)$, $(7,0)$, $(2,5)$ and $(0,4)$, as shown in Figure 3.5, whose interior is R and whose boundary is S. Set $(\bar{x},\bar{y}) = (0,0)$ and $h = 2$. Let S_1, S_2, S_3, and S_4 denote the four sides of the quadrilateral, as shown in Figure 3.5. Then the points of R_h are $(2,2)$, $(2,4)$ and $(4,2)$ and have been crossed in Figure 3.5. The points of S_h^* are all of the points in S_1 and S_2 and the four circled and one squared point of S_3. The points of S_h are $(2,0)$, $(4,0)$, $(0,2)$, $(5,2)$, $(4,3)$, $(0,4)$, $(3,4)$ and $(2,5)$, which are circled in the figure.

3.5 The Numerical Method

We formulate now the basic algorithm for approximating the solution of the Dirichlet problem.

Method D

For fixed $h > 0$ and fixed $(\bar{x},\bar{y})$, construct R_h and S_h. Suppose R_h consists of m points and S_h consists of n points. Number the points of R_h in a one-to-one fashion with the integers $1 - m$ in such a way that the numbers are increasing from left to

NUMERICAL METHOD

right on any horizontal line of the lattice and increasing from bottom to top on any vertical line of the lattice. Number the points of S_h in a one-to-one fashion, and in any order, with the integers $m+1$, $m+2, \ldots, m+n$.

Step 1. At each point of S_h, set

$$u(x,y) = f(x,y).$$

If (x,y) is numbered k, then this is equivalent, in subscript notation, to

$$u_k = f(x,y).$$

Step 2. At each point (x,y) of R_h, beginning with the one numbered 1 and continuing consecutively through the one numbered m, write down the Laplace difference analogue

(3.17)
$$-2\left(\frac{1}{h_1 h_3} + \frac{1}{h_2 h_4}\right) u(x,y) + \frac{2}{h_1(h_1+h_3)} u(x+h_1, y)$$
$$+ \frac{2}{h_2(h_2+h_4)} u(x, y+h_2) + \frac{2}{h_3(h_1+h_3)} u(x-h_3, y)$$
$$+ \frac{2}{h_4(h_2+h_4)} u(x, y-h_4) = 0,$$

where $(x+h_1, y)$, $(x, y+h_2)$, $(x-h_3, y)$, $(x, y-h_4)$ are the neighbors of (x,y). In so doing, if any neighbor is a point of S_h, then replace the corresponding u value by the known value of f determined in

Step 1. In practice, each equation should be written in subscript notation, as demonstrated in (3.16), so that there results a linear algebraic system of m equations in the m unknowns $u_1, u_2, \ldots, u_m$.

Step 3.   Solve the algebraic system generated in Step 2.

Step 4.   Let the discrete function u_i, $i = 1, 2, \ldots, m+n$, which is defined only on $R_h + S_h$, represent on $R_h + S_h$ the approximate solution of the given Dirichlet problem.

Example

Let S be the quadrilateral with vertices $(0,0)$, $(7,0)$, $(2,5)$ and $(0,4)$, which is shown in Figure 3.5. Let R be the interior of S. On $R + S$ consider the Dirichlet problem with $f(x,y) = x^2 - y^2$. Set $(\bar{x}, \bar{y}) = (0,0)$ and $h = 2$, as in the previous example. As shown in Figure 3.6, the points of R_h are numbered 1-3 while those of S_h are numbered 4-11. Following the directions of Step 1, one has

(3.18) $u_4 = 4$, $u_5 = 16$, $u_6 = -4$, $u_7 = 21$, $u_8 = 7$, $u_9 = -16$,

$$u_{10} = -7, \quad u_{11} = -21.$$

Application of (3.17) at the points numbered 1-3 in Figure 3.6 and substitution from (3.18) yields

$$(-1)u_1 + \frac{1}{4}u_2 + \frac{1}{4}u_3 + \frac{1}{4}(-4) + \frac{1}{4}(4) = 0$$

$$(-2)u_2 + \frac{2}{1(1+2)}(21) + \frac{2}{1(1+2)}(7) + \frac{2}{2(1+2)}u_1 + \frac{2}{2(1+2)}(16) = 0$$

$$(-2)u_3 + \frac{2}{1(1+2)}(-7) + \frac{2}{1(1+2)}(-21) + \frac{2}{2(1+2)}(-16) + \frac{2}{2(1+2)}u_1 = 0,$$

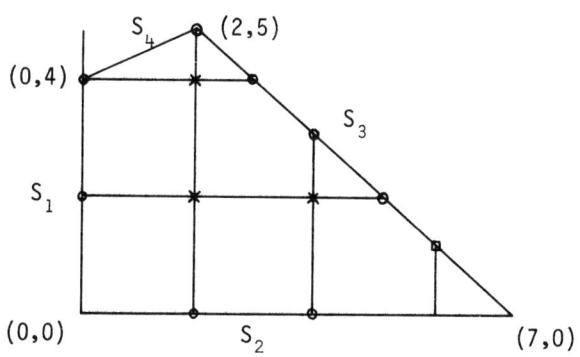

Figure 3.5

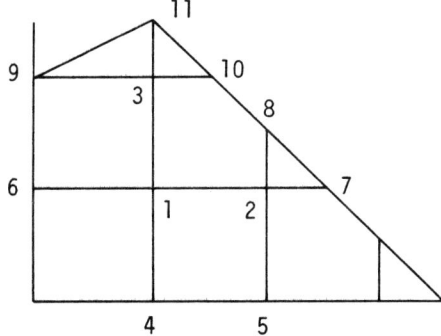

Figure 3.6

or, equivalently,

(3.19) $\quad -u_1 + \frac{1}{4}u_2 + \frac{1}{4}u_3 = 0$

(3.20) $\quad \frac{1}{3}u_1 - 2u_2 = -24$

(3.21) $\quad \frac{1}{3}u_1 - 2u_3 = 24.$

The solution of (3.19)–(3.21) is

(3.22) $\quad u_1 = 0, \; u_2 = 12, \; u_3 = -12.$

Thus, u_i, $i = 1, 2, \ldots, 11$, as given by (3.18) and (3.22), constitutes the approximate solution of the given Dirichlet problem on $R_h + S_h$.

Observe that the significance of the ordering in Step 3 of Method D is that the linear algebraic system which results is diagonally dominant, since the main diagonal terms come from the coefficient of $u(x,y)$ in (3.17).

The reasonableness of Method D as a numerical method follows from the known results (Greenspan (3)) that (a) the approximate solution always exists and is unique, (b) for a large class of problems the numerical solution converges to the analytical solution as the grid size converges to zero, and (c) the system of algebraic equations generated by Method D can be solved by SOR, with convergence assured for any initial guess and for any ω in the range $0 < \omega < 2$.

EXTERIOR PROBLEMS

Moreover, for certain classes of problems, one can even calculate the value of ω which will yield the maximal rate of convergence for the SOR method (Warlick and Young). Thus, for example, if $a \geq b > 0$, and if S is a rectangle with vertices $(0,0)$, $(a,0)$, $(0,b)$, and (a,b), then

$$\omega = \frac{2}{1 + \sqrt{1 - \lambda^2}},$$

where

$$\lambda = \frac{1}{2}\left(\cos\frac{\pi}{A} + \cos\frac{\pi}{B}\right)$$

and

$$a = Ah, \quad b = Bh.$$

3.6 Numerical Solution of the Exterior Dirichlet Problem

Numerically, the exterior Dirichlet problem also can be solved easily if one first transforms it into an equivalent interior problem and then applies Method D. This can be done as follows.

For simplicity, let C be a circle whose center is $(0,0)$ and whose radius is unity. Let L be any half-line which emanates from the origin (see Figure 3.7). If $P(x,y)$ is any point on L which is different from the origin, then the unique point $Q(\xi,\eta)$ on L for which

(3.23) $$|OP| \cdot |OQ| = 1$$

is called the inverse point of P. The mapping of all points of the plane, other than the origin, into their inverse points, is called an inversion mapping. In effect, points inside C map into points outside C, points on C map into themselves, and points outside C map into points inside C. Thus, any unbounded set outside C maps into a bounded set inside C.

The equations of inversion mapping can be developed easily as follows. As shown in Figure 3.7, let the foot of the perpendicular to the X axis through P be P' and that through Q be Q'. Then, by similar triangles,

(3.24) $$\frac{x}{\sqrt{x^2+y^2}} = \frac{\xi}{\sqrt{\xi^2+\eta^2}}.$$

From (3.23)

(3.25) $$\sqrt{x^2+y^2} \cdot \sqrt{\xi^2+\eta^2} = 1,$$

so that

(3.26) $$\xi = \frac{x}{x^2+y^2}, \quad x^2+y^2 \neq 0.$$

Similarly, by constructing perpendiculars to the Y axis,

(3.27) $$\eta = \frac{y}{x^2+y^2}, \quad x^2+y^2 \neq 0.$$

EXTERIOR PROBLEMS   93

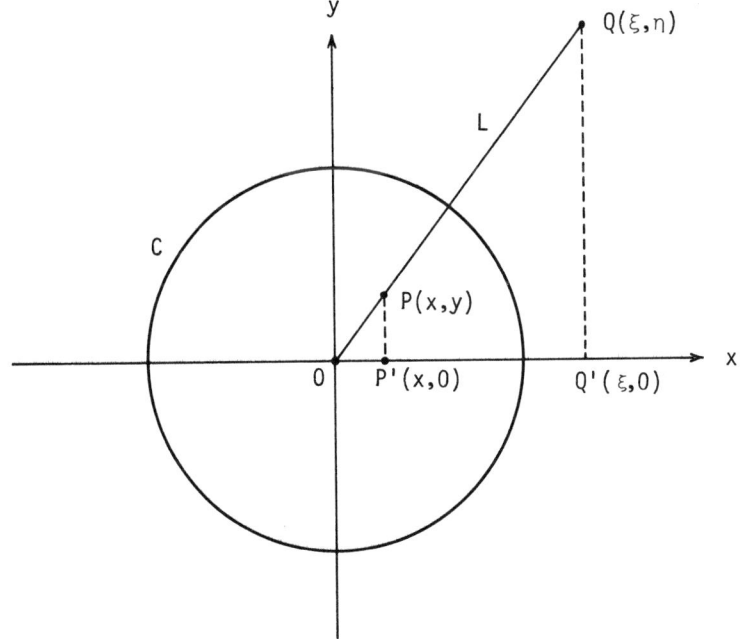

Figure 3.7

Formulas (3.26) and (3.27) are convenient for determining ξ and η when x and y are given. For transforming a given equation in x and y to one in ξ and η, it is more convenient to have (3.26) and (3.27) solved for x and y in terms of ξ and η. These formulas can be written in the form

(3.28) $$x = \frac{\xi}{\xi^2 + \eta^2}, \quad \xi^2 + \eta^2 \neq 0$$

(3.29) $$y = \frac{\eta}{\xi^2+\eta^2}, \quad \xi^2+\eta^2 \neq 0.$$

Consider now the following well known theorem (Petrovsky).

Theorem 3.1

Let $u(x,y)$ be the solution of the exterior Dirichlet problem. Without loss of generality, assume that $(0,0)$ is in R. Under inversion, let

(3.30) $$R^* \to R^i, \quad S \to S^i$$

(3.31) $$u(x,y) = u\left(\frac{\xi}{\xi^2+\eta^2}, \frac{\eta}{\xi^2+\eta^2}\right) = v(\xi,\eta)$$

(3.32) $$f(x,y) = f\left(\frac{\xi}{\xi^2+\eta^2}, \frac{\eta}{\xi^2+\eta^2}\right) = F(\xi,\eta).$$

Then $v(\xi,\eta)$ is the solution of the Dirichlet problem on $R^i + S^i$ with boundary values F, that is,

(a) $\frac{\partial^2 v}{\partial \xi^2} + \frac{\partial^2 v}{\partial \eta^2} = 0, \quad (\xi,\eta)$ in R^i

(b) $v(\xi,\eta)$ is defined and continuous on $R^i + S^i$, and

(c) $v(\xi,\eta) = F(\xi,\eta),$ on $S^i.$

The value of Theorem 3.1 is that it enables us to apply Method D to a Dirichlet problem for v and then to determine approximate values for u, the solution of the exterior problem, directly from (3.31).

GENERAL LINEAR EQUATIONS

3.7 Remark on Neumann and Mixed Type Problems

Method D extends easily to mixed type problems (Greenspan (3)), but in general with less accuracy. However, rather than introduce the pertinent new ideas now, we shall do so when considering problems in which normal derivative boundary conditions are natural. We will not attempt to deal with Neumann problems numerically because they are not well posed. If one does have a Neumann problem, however, it is important to note that prescribing the solution at only one boundary point transforms the problem into one which is well posed.

3.8 The General Linear Elliptic Equation with Constant Coefficients

In this section, let us consider those modifications of Method D which are necessary when the Laplace equation is replaced by a different linear elliptic equation. The discussion will focus on equations which occur repeatedly in physical applications.

If A, B, C, D, E, F are constants and $G(x,y)$ is continuous, it is known (Courant and Hilbert, Greenspan (2)) that the partial differential equation

$$(3.33) \quad Au_{xx} + 2Bu_{xy} + Cu_{yy} + Du_x + Eu_y + Fu + G(x,y) = 0,$$
$$A^2 + B^2 + C^2 \neq 0$$

can be simplified by a rotation of axes. When (3.33) is elliptic, one can, for example, eliminate the u_{xy} term. Thus, without loss of generality, let us assume that $B = 0$. Consider, then,

(3.34) $\quad\quad Au_{xx} + Cu_{yy} + Du_x + Eu_y + Fu + G(x,y) = 0$

and assume that

(3.35) $\quad\quad\quad\quad A > 0, \quad C > 0$,

so that the equation is elliptic. Further, for both practical and theoretical reasons, it will be convenient at present to assume that

(3.36) $\quad\quad\quad\quad\quad F \leq 0$,

which will assure (Courant and Hilbert) that any solution of (3.34) has certain properties, like a general max-min property, in common with harmonic functions.

To construct a difference approximation of (3.34), consider the five-point arrangement shown in Figure 3.3 and at (x,y) set

(3.37) $\quad Au_{xx} + Cu_{yy} + Du_x + Eu_y + Fu + G(x,y) \equiv \sum_{0}^{4} \alpha_i u_i + G(x,y)$.

Substitution of finite Taylor expansions about (x,y) into (3.37) and setting corresponding coefficients equal yields the system

$$\alpha_0 + \alpha_1 + \alpha_2 + \alpha_3 + \alpha_4 = F$$

GENERAL LINEAR EQUATIONS

$$h_1\alpha_1 - h_3\alpha_3 = D, \quad h_2\alpha_2 - h_4\alpha_4 = E$$
$$h_1^2\alpha_1 + h_3^2\alpha_3 = 2A, \quad h_2^2\alpha_2 + h_4^2\alpha_4 = 2C,$$

the unique solution of which is

(3.38) $$\alpha_0 = F - \frac{2A}{h_1 h_3} - \frac{2C}{h_2 h_4} - \frac{D(h_3-h_1)}{h_1 h_3} - \frac{E(h_4-h_2)}{h_2 h_4}$$

$$\alpha_1 = \frac{2A + Dh_3}{h_1(h_1+h_3)}, \quad \alpha_2 = \frac{2C + Eh_4}{h_2(h_2+h_4)},$$

$$\alpha_3 = \frac{2A - Dh_1}{h_3(h_1+h_3)}, \quad \alpha_4 = \frac{2C - Eh_2}{h_4(h_2+h_4)}.$$

Because the truncation error goes to zero with h, the difference equation approximation of (3.34) is chosen to be

(3.39) $$\sum_{0}^{4} \alpha_i u_i + G(x,y) = 0,$$

where the α_i are given by (3.38).

If one has a Dirichlet problem in which the Laplace equation is replaced by (3.34), then Method D need be modified only by replacing (3.17) with (3.39). However, if one wishes assurance, a priori, that the theoretical support available for Method D is also available for the modified method, then (Greenspan (3)) one need only select h small enough so that, in (3.38), one has

(3.40) $$\alpha_0 < 0, \quad \alpha_i > 0, \quad i = 1,2,3,4.$$

From (3.38), a sufficient condition for (3.40) to be valid is

(3.41) $$\max[\,|D|/A,\;|E|/C\,] < \frac{2}{h}.$$

Note that one can also apply the forward-backward technique of Section 2.8 to develop a difference analogue which, in general, is less accurate, but which yields diagonal dominance for any grid size. This will, in fact, be done in Chapter VII.

3.9 Extension to Three Dimensions

The numerical analysis developed thus far generalizes easily and naturally to linear problems in any number of dimensions. For clarity, however, we shall give a detailed discussion and a significant application only for three dimensional problems for the Laplace equation.

Let R be a bounded, three dimensional region and let S be its boundary. Let $f(x,y,z)$ be defined and continuous on S. Then the Dirichlet problem is that of finding a function $u(x,y,z)$ such that

(a) u satisfies on R the Laplace equation
$$u_{xx} + u_{yy} + u_{zz} = 0,$$

(b) $u = f$ on S, and

(c) u is continuous on R + S.

THREE DIMENSIONAL PROBLEMS

Under several reasonable assumptions about S (Petrovsky), which, though quite general, are somewhat more restrictive than those for the two dimensional case, it is known that the Dirichlet problem has a unique solution. As in Method D, in order to approximate this solution one need only construct, for $h > 0$, three dimensional, finite point sets R_h (interior grid points) in R and S_h (boundary grid points) in S, where in three dimensions each point of R_h has six neighbors (see Figure 3.8), and then solve the linear algebraic system which results by applying at each point (x,y,z) of R_h (see the notation in Figure 3.8) the difference equation

$$(3.42) \quad -2\left(\frac{1}{h_1 h_2} + \frac{1}{h_3 h_4} + \frac{1}{h_5 h_6}\right) u_0 + \frac{2}{h_1(h_1+h_2)} u_1 + \frac{2}{h_2(h_1+h_2)} u_2$$

$$+ \frac{2}{h_3(h_3+h_4)} u_3 + \frac{2}{h_4(h_3+h_4)} u_4 + \frac{2}{h_5(h_5+h_6)} u_5$$

$$+ \frac{2}{h_6(h_5+h_6)} u_6 = 0.$$

Note that (3.42) is a natural extension of (3.16) and that (3.42) can be developed in the same fashion as was (3.16).

The exterior Dirichlet problem can be formulated as follows. Let R be a bounded, three dimensional region and let S be its boundary. Let R^* be the exterior of S and let $f(x,y,z)$ be defined and continuous on S. Then the exterior Dirichlet problem is that of

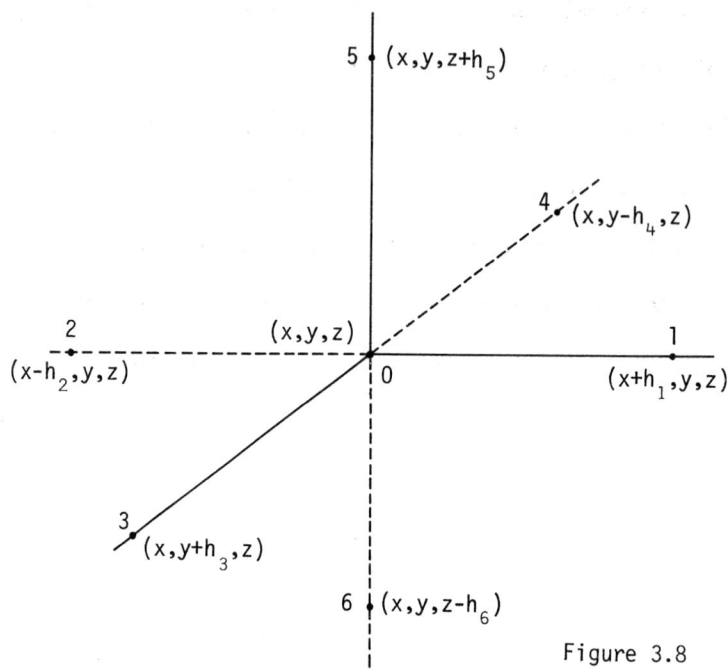

Figure 3.8

finding a function $u(x,y,z)$ such that

(a) u satisfies, on R^*, the Laplace equation

$$u_{xx} + u_{yy} + u_{zz} = 0,$$

(b) $u = f$ on S,

(c) u is continuous on $R^* + S$, and

(d) u is bounded on $R^* + S$.

Again (Courant and Hilbert), it is known that, under rather general restrictions on S, the exterior Dirichlet problem has a unique solution.

THREE DIMENSIONAL PROBLEMS

Unfortunately, the method of Section 3.6, for transforming an exterior problem into an interior problem, does <u>not</u> extend, per se, in three dimensions. With the following simple modification, however, it will extend.

As in (3.30)-(3.31), inversion with respect to a unit sphere in three dimensions is given by

(3.43) $\quad x = \dfrac{\xi}{\xi^2+\eta^2+\nu^2}, \; y = \dfrac{\eta}{\xi^2+\eta^2+\nu^2}, \; z = \dfrac{\nu}{\xi^2+\eta^2+\nu^2} \; ; \; \xi^2+\eta^2+\nu^2 \neq 0,$

or, equivalently, by

(3.44) $\quad \xi = \dfrac{x}{x^2+y^2+z^2}, \; \eta = \dfrac{y}{x^2+y^2+z^2}, \; \nu = \dfrac{z}{x^2+y^2+z^2} \; ; \; x^2+y^2+z^2 \neq 0.$

Consider now the following theorem (Petrovsky).

Theorem 3.2

Let $u(x,y,z)$ be the solution of the exterior Dirichlet problem. Without loss of generality, assume that $(0,0,0)$ is in R. Under inversion, let $R^* \to R^i$, $S \to S^i$, and

(3.45) $\quad u(x,y,z) = u(\dfrac{\xi}{\xi^2+\eta^2+\nu^2}, \dfrac{\eta}{\xi^2+\eta^2+\nu^2}, \dfrac{\nu}{\xi^2+\eta^2+\nu^2}) = V(\xi,\eta,\nu)$

(3.46) $\quad f(x,y,z) = f(\dfrac{\xi}{\xi^2+\eta^2+\nu^2}, \dfrac{\eta}{\xi^2+\eta^2+\nu^2}, \dfrac{\nu}{\xi^2+\eta^2+\nu^2}) = \mathfrak{F}(\xi,\eta,\nu).$

Define $v(\xi,\eta,\nu)$ and $F(\xi,\eta,\nu)$ by

(3.47) $\quad v(\xi,\eta,\nu) = [V(\xi,\eta,\nu)] / \sqrt{\xi^2 + \eta^2 + \nu^2}$

(3.48) $\quad F(\xi,\eta,\nu) = [\mathfrak{F}(\xi,\eta,\nu)] / \sqrt{\xi^2 + \eta^2 + \nu^2}$.

Then $v(\xi,\eta,\nu)$ is the solution of the Dirichlet problem on $R^i + S^i$ with boundary function F, that is

(a) $\quad v_{\xi\xi} + v_{\eta\eta} + v_{\nu\nu} = 0$, on R^i

(b) $\quad v(\xi,\eta,\nu)$ is defined and continuous on $R^i + S^i$, and

(c) $\quad v(\xi,\eta,\nu) = F(\xi,\eta,\nu)$, on S^i.

With regard to solving the three dimensional, exterior Dirichlet problem numerically by first applying an inversion mapping, Theorem 3.2 implies that u, v, f and F are related by

(3.49) $\quad u(x,y,z) = [\sqrt{\xi^2 + \eta^2 + \nu^2}\,] \, v(\xi,\eta,\nu)$

(3.50) $\quad f(x,y,z) = [\sqrt{\xi^2 + \eta^2 + \nu^2}\,] \, F(\xi,\eta,\nu)$,

which are, indeed, different in character from the two-dimensional relationships (3.31) and (3.32).

3.10 The Classical Problem of Capacity

Rather than merely give a trivial illustrative example of the numerical solution of a three dimensional problem, let us consider a physical problem which is of long standing interest, which is exceptionally difficult to solve analytically, and which has applications in such diverse areas as electron optics, antenna design, plasma dynamics, and electrostatics, that is, the problem of capacity.

In the exterior Dirichlet problem, if one sets

$$f(x,y,z) = 1$$

and if $\frac{\partial u}{\partial n}$ is the outward normal derivative on S of the solution of the resulting problem, then the capacity C of S is defined by the surface integral

$$(3.51) \qquad C = -\frac{1}{4\pi} \iint_S \frac{\partial u}{\partial n} \, dA \, .$$

From, say, the electrostatic point of view, the capacity C of S is the total charge which, in equilibrium on S, raises the surface potential to unity.

Unfortunately, for any nonspherical surface, the exact value of C is, in general, so difficult to determine that even the capacity of the unit cube has become a quantity of great interest. Mathematicians have approached such problems by means of isoperimetric inequalities, while physicists and engineers have been prone to apply infinite series techniques. The isoperimetric inequality approach requires special results for each S and yields upper and lower bounds for C which are rarely sharp. The infinite series approach usually requires extensive tables, which are _different_ for each S, and which may have to be so voluminous to attain reasonable accuracy, that the method becomes impractical.

We shall show next, then, how to apply our numerical method in a completely general and efficient way to estimate the capacity of _any_ surface. The key to the method lies in the known result (Greenspan (6)) that if u and v are related by (3.49), then C can be given also by

(3.52) $$C = v(0,0,0).$$

For illustrative purposes, let us show how to calculate the capacity of a unit cube. Let S be the cube whose vertices are (1/2, 1/2, 1/2), (1/2, 1/2, -1/2), (1/2, -1/2, 1/2), (-1/2, 1/2, 1/2), (1/2, -1/2, -1/2), (-1/2, 1/2, -1/2), (-1/2, -1/2, 1/2), (-1/2, -1/2, -1/2), as shown in Figure 3.9. Then S^i, the map of S under inversion transformation (3.43) or (3.44), is a completely symmetrical surface consisting of six partially spherical caps, the first octant of which is shown in Figure 3.10. With boundary function $F(\xi, \eta, \nu) = (\xi^2 + \eta^2 + \nu^2)^{-1/2}$, the numerical method of Section 3.9 was applied with grid size $h = 0.045$ and with $\omega = 1.94$ on the UNIVAC 1108 to yield, in only four minutes, the approximation

CAPACITY   105

$$C = v(0,0,0) = 0.661.$$

By means of isoperimetric inequalities the following upper and lower bounds have been obtained after some forty years of research (Polya and Szego, Greenspan (6))

$$0.632 < C < 0.6626.$$

For the calculations of the capacities of ellipsoids, lenses and toroids, see Greenspan (6).

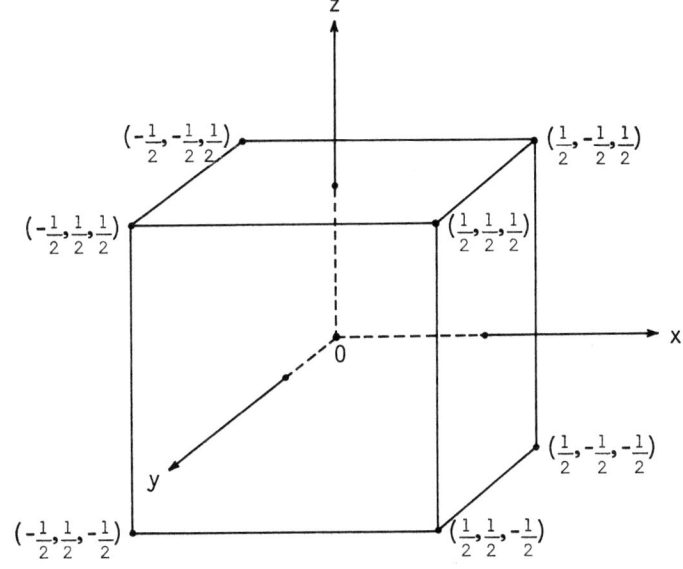

Figure 3.9

106   ELLIPTIC EQUATIONS

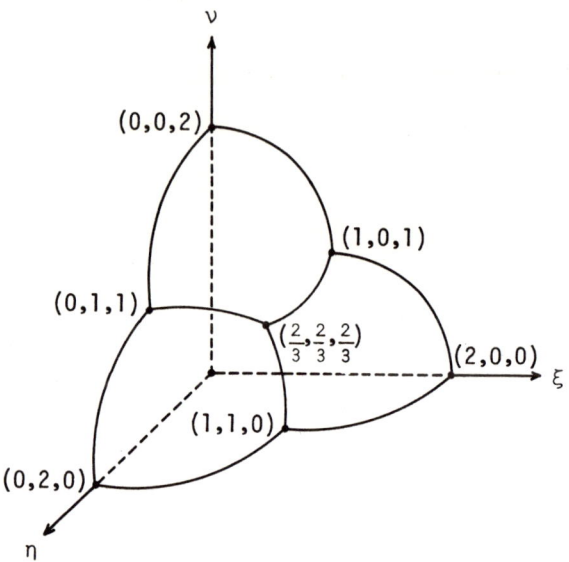

Figure 3.10

3.11 Mildly Nonlinear Problems

We return now to two dimensional problems, but begin the study of nonlinear equations. The three prototype problems of the classes of elliptic equations to be considered are

MILDLY NONLINEAR PROBLEMS

(3.53) $\quad u_{xx} + u_{yy} = e^u \quad$ (Radiation equation)

(3.54) $\quad u_{xx} + u_{yy} = u^2 \quad$ (Molecular interaction equation)

(3.55) $\quad (1+u_y^2)u_{xx} - 2u_x u_y u_{xy} + (1+u_x^2)u_{yy} = 0 \quad$ (Soap film equation)

Equations (3.53) and (3.54) are mildly nonlinear and will be studied in this section. Study of equation (3.55) will have to be deferred until Chapter 6.

Consider then the mildly nonlinear elliptic equation

(3.56) $\quad\quad u_{xx} + u_{yy} = F(x,y,u)$.

We will assume that

(3.57) $\quad\quad \dfrac{\partial F}{\partial u} \geq 0$,

in order to be assured that solutions of the Dirichlet problem for (3.56) exist, are unique, and have certain general properties in common with harmonic functions (Courant and Hilbert). (Note immediately that (3.53) satisfies (3.57) but (3.54) does not.) Method D now need be modified only by replacing linear difference equation (3.16) with nonlinear difference equation

(3.58) $\quad -2\left[\dfrac{1}{h_1 h_3} + \dfrac{1}{h_2 h_4}\right] u_0 + \dfrac{2}{h_1(h_1+h_3)} u_1 + \dfrac{2}{h_2(h_2+h_4)} u_2$

$\quad\quad + \dfrac{2}{h_3(h_1+h_3)} u_3 + \dfrac{2}{h_4(h_2+h_4)} u_4 = F(x,y,u_0)$, and the resulting numerical method is mathematically respectable.

Example

Let S be the square with vertices $(0,0)$, $(1,0)$, $(1,1)$, $(0,1)$ and let R be its interior. On S, set $f(x,y) = 0$. For $h = 1/3$, the points of R_h, as shown in Figure 3.11, are numbered 1, 2, 3, 4. If the differential equation defined on R is (3.53), then application of (3.58) at each point of R_h yields the nonlinear system

$$e^{u_1} - 36u_1 + 9u_2 + 9u_3 = 0$$
$$9u_1 - e^{u_2} - 36u_2 + 9u_4 = 0$$
$$9u_1 - e^{u_3} - 36u_3 + 9u_4 = 0$$
$$9u_2 + 9u_3 - e^{u_4} - 36u_4 = 0.$$

This system can be solved easily by the generalized Newton's method to yield the numerical approximation. For more extensive examples, see Greenspan (6).

It should be noted also that the method outlined for (3.56) extends, but with several additional assumptions (Bers), to

$$u_{xx} + u_{yy} = F(x, y, u, u_x, u_y).$$

MILDLY NONLINEAR PROBLEMS

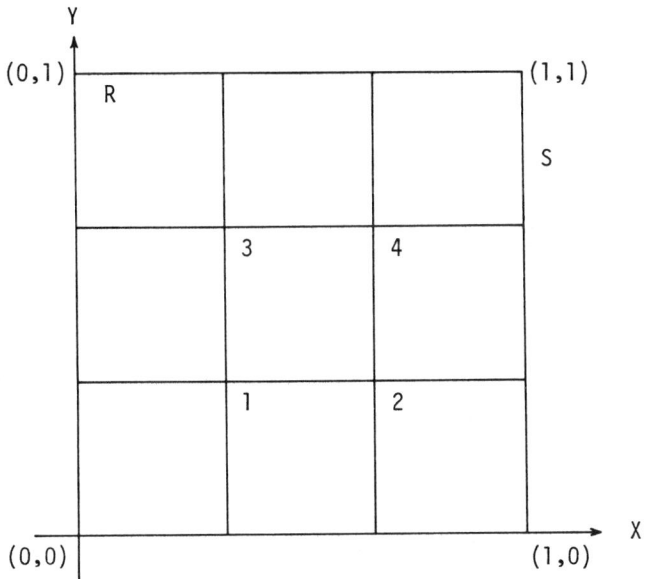

Figure 3.11

Since equation (3.54) does not satisfy (3.57), we will devise an alternative method for it. This new method, incidentally, will also be applicable to (3.53), but with less efficiency than the method already devised. It is important to note, first, however, that the Dirichlet problem for (3.54) need not have a unique solution. In order to consider a problem which does have a unique solution, physical considerations lead to the requirement that the boundary function f(x,y) be non-negative on S. It follows then (Pohozaev) that the Dirichlet problem for (3.54) has a unique nonnegative solution. Our attention, then will be directed toward solving the Dirichlet problem for mildly non-linear equation (3.56) subject to the conditions

(3.59) $\qquad f(x,y) \geq 0 \text{ on } S$

and

(3.60) $\qquad \frac{\partial F}{\partial u} \geq 0 \text{ if } u \geq 0.$

Such problems have unique non-negative solutions which can be approximated by discretizing Pohozaev's analytical method, in which he first reformulates the problem as an integral equation and then applies a Banach space form of Newton's method to solve the resulting integral equation iteratively. The Pohozaev analytical iteration formula for (3.56) is

(3.61) $\quad \Delta u^{(n+1)} - F_u(x,y,u^{(n)})u^{(n+1)} = F(x,y,u^{(n)}) - F_u(x,y,u^{(n)})u^{(n)},$

$\qquad n = 0, 1, \ldots,$

which, one should observe, represents a sequence of linear elliptic equations in $u^{(n+1)}$. The precise details of the method are now given by means of an illustrative example.

Example

Let S be the square with vertices (0,0), (1,0), (1,1) and (0,1), and let R be the interior of S. On S define

MILDLY NONLINEAR PROBLEMS

$$f(x,y) = 1$$

and consider the Dirichlet problem for

$$\Delta u = u^2 .$$

For this equation, (3.61) takes the form

(3.62) $\quad \Delta u^{(n+1)} - 2u^{(n)}u^{(n+1)} = -[u^{(n)}]^2, \quad n = 0,1,2,\ldots .$

In terms of the point arrangement shown in Figure 3.3, a discretized form of (3.62) is

(3.63) $\quad -2\left[\dfrac{1}{h_1 h_3} + \dfrac{1}{h_2 h_4}\right]u_0^{(n+1)} + \dfrac{2}{h_1(h_1+h_3)}u_1^{(n+1)} + \dfrac{2}{h_2(h_2+h_4)}u_2^{(n+1)}$

$\qquad + \dfrac{2}{h_3(h_1+h_3)}u_3^{(n+1)} + \dfrac{2}{h_4(h_2+h_4)}u_4^{(n+1)} - 2u_0^{(n)}u_0^{(n+1)}$

$\qquad = -[u_0^{(n)}]^2, \quad n = 0,1,2,\ldots .$

For $h = \dfrac{1}{3}$, (3.63) reduces to

(3.64) $\quad -(36 + 2u_0^{(n)})u_0^{(n+1)} + 9u_1^{(n+1)} + 9u_2^{(n+1)} + 9u_3^{(n+1)} + 9u_4^{(n+1)}$

$\qquad = -[u_0^{(n)}]^2, \quad n = 0,1,2,\ldots .$

Next, the points of R_h are numbered as in Figure 3.11.

Now (3.64) can be used as an iterative formula only if $u_1^{(0)}$, $u_2^{(0)}$, $u_3^{(0)}$ and $u_4^{(0)}$ are given. We choose these to be the numerical solution of the Dirichlet problem for the Laplace equation on $R + S$ with the given f. Thus, $u_1^{(0)}$, $u_2^{(0)}$, $u_3^{(0)}$ and $u_4^{(0)}$ are determined by Method D, and, in this case, turn out to be

(3.65) $$u_1^{(0)} = u_2^{(0)} = u_3^{(0)} = u_4^{(0)} = 1 .$$

Next, one applies (3.64) with $n = 0$ at each point of R_h to yield, with the aid of (3.65), the four equations

$$-38u_1^{(1)} + 9u_2^{(1)} + 9u_3^{(1)} = -19$$
$$-38u_2^{(1)} + 9u_4^{(1)} + 9u_1^{(1)} = -19$$
$$-38u_3^{(1)} + 9u_4^{(1)} + 9u_1^{(1)} = -19$$
$$-38u_4^{(1)} + 9u_3^{(1)} + 9u_2^{(1)} = -19 .$$

The solution of this system by SOR yields $u_1^{(1)}$, $u_2^{(1)}$, $u_3^{(1)}$, $u_4^{(1)}$. Knowing these, one proceeds to apply (3.54) with $n = 1$ at each point of R_h to yield the system

$$-(36+2u_1^{(1)})u_1^{(2)} + 9u_2^{(2)} + 9u_3^{(2)} + 9 + 9 = -[u_1^{(1)}]^2$$
$$-(36+2u_2^{(1)})u_2^{(2)} + 9 + 9u_4^{(2)} + 9u_1^{(2)} + 9 = -[u_2^{(1)}]^2$$
$$-(36+2u_3^{(1)})u_3^{(2)} + 9u_4^{(2)} + 9 + 9 + 9u_1^{(2)} = -[u_3^{(1)}]^2$$
$$-(36+2u_4^{(1)})u_4^{(2)} + 9 + 9 + 9u_3^{(2)} + 9u_2^{(2)} = -[u_4^{(1)}]^2$$

which, when solved by SOR, yields $u_1^{(2)}$, $u_2^{(2)}$, $u_3^{(2)}$, $u_4^{(2)}$. In the indicated fashion, the iteration continues until, for some value k, one has

$$u_1^{(k)} = u_1^{(k+1)}, \ u_2^{(k)} = u_2^{(k+1)}, \ u_3^{(k)} = u_3^{(k+1)}, \ u_4^{(k)} = u_4^{(k+1)},$$

and the approximate solution is taken to be $u_1^{(k)}, u_2^{(k)}, u_3^{(k)}, u_4^{(k)}$.

It is worth noting, finally, that the above method, based on solving a sequence of linear problems, has a firm mathematical basis (Greenspan (3)), and that the technique of studying a nonlinear equation as a sequence of linear equations can be of exceptional value.

Exercises

1. Classify each of the following partial differential equations as elliptic, parabolic, or hyperbolic at the point $(0,0)$.

   (a) $u_{xx} + 2u_{yy} = 0$

   (b) $u_{xx} - 2u_{yy} = 0$

   (c) $u_{xx} - 2u_y = 0$

   (d) $u_{xx} - 4u_{xy} + u_{yy} = 0$

   (e) $3u_{xx} - 4u_{xy} - 5u_{yy} = 0$

   (f) $3u_{xx} - 4u_{xy} - 5u_{yy} + 8u_x - 9u_y + 6u = 27e^{xy}$.

2. Determine, if possible, in which portions of the plane each of the following is elliptic, parabolic, and hyperbolic.

   (a) $yu_{xx} - u_{yy} = 0$

   (b) $u_{xx} + 2xu_{xy} + (1-y^2)u_{yy} = 0$

   (c) $(1-u_x^2)u_{xx} - 2u_x u_y u_{xy} + (1-u_y^2)u_{yy} - 8u_x = e^u$.

3. Let S be the square whose vertices are $(1/2, 1/2)$, $(-1/2, 1/2)$, $(-1/2, -1/2)$ and $(1/2, -1/2)$ and let R be the interior of S. Show that each of the following functions is continuous on $R+S$, harmonic on R, and takes on its maximum and minimum values on S.

   (a) $u = 5$ (b) $u = 4y - 7$ (c) $u = 7x - 4y - 2$

   (d) $u = x^2 - y^2$ (e) $u = \frac{1}{2}xy^2 - \frac{x^3}{6}$ (f) $u = \frac{xy^3 - x^3y}{6}$.

EXERCISES

4. Repeat Exercise 3 but let S be the unit circle, whose equation is $x^2 + y^2 = 1$.

5. With $h = 2$ and $(\bar{x}, \bar{y}) = (0,0)$, find the numerical solution of the Dirichlet problem for which $f(x,y) = x - 2y$ and S is the triangle whose vertices are $(0,0)$, $(7,0)$ and $(0,7)$.

6. With $h = 2$ and $(\bar{x}, \bar{y}) = (0,0)$, find the numerical solution of the Dirichlet problem for which $f(x,y) = x^2 - y^2$ and S is the rectangle whose vertices are $(0,0)$, $(5,0)$, $(5,4)$, $(0,4)$.

7. With $h = 1/2$ and $(\bar{x}, \bar{y}) = (0,0)$, find the numerical solution of the Dirichlet problem for which $f(x,y) = x^2 - y$ and S is the circle of unit radius whose center is $(1,1)$.

8. Let S be the square whose vertices are $(1,1)$, $(2,1)$, $(2,2)$, $(1,2)$ and let R be the interior of S. Let $f(x,y) = x - y^3$ on S and consider the resulting Dirichlet problem. By a change of variables, transform the problem into one in which the origin lies interior to the region of interest.

9. In each of the following, transform the associated exterior Dirichlet problem into an equivalent interior problem.
   (a) S is the circle whose equation is $x^2 + y^2 = 1$, $f(x,y) = 1$
   (b) S is the circle whose equation is $(x-2)^2 + y^2 = 1$, $f(x,y) = xy^2$

(c) S is the ellipse whose equation is $x^2+4y^2=1$, $f(x,y) = x+y^2$

(d) S is the square whose vertices are $(1/2,1/2)$, $(1/2,-1/2)$, $(-1/2,1/2)$, $(-1/2,-1/2)$; $f(x,y) = \dfrac{x^2-y^2}{(x^2+y^2)^2}$

(e) S is the rectangle whose vertices are $(-1,-1)$, $(4,-1)$, $(-1,6)$ $(4,6)$; $f(x,y) = \sin(x+y)$.

10. Solve exterior problem 9(d) numerically and compare your results with those of the exact solution $u = \dfrac{x^2-y^2}{(x^2+y^2)^2}$.

11. Prove that if (a) $h \geq h_i > 0$, $i = 1,2,3,4$, (b) $A > 0$, $C > 0$, $F \leq 0$, (c) $\max[\,|D|/A,\ |E|/C\,] < 2/h$, then

$$F - \frac{2A}{h_1 h_3} - \frac{2C}{h_2 h_4} - \frac{D(h_3-h_1)}{h_1 h_3} - \frac{E(h_2-h_4)}{h_2 h_4} < 0.$$

12. If S is the unit cube whose vertices are $(0,0,0)$, $(1,0,0)$, $(0,1,0)$, $(0,0,1)$, $(1,1,0)$, $(1,0,1)$, $(0,1,1)$, $(1,1,1)$, and if $f(x,y,z) = xy+z^2$ on S, then find the numerical solution of the resulting Dirichlet problem using $(\bar{x},\bar{y},\bar{z}) = (0,0,0)$ and $h = 2/5$.

13. Find a relationship between the capacity and the radius of an arbitrary sphere.

14. Find the capacity of the ellipsoid whose equation is

$$\frac{x^2}{9} + \frac{y^2}{4} + z^2 = 1.$$

EXERCISES

15. Let S be the unit square whose vertices are $(1/2,1/2)$, $(1/2,-1/2)$, $(-1/2,1/2)$, $(-1/2,-1/2)$, let R be the interior of S, and let $f(x,y) = 1$ on S. For $h = 1/3$, find a numerical solution of the resulting Dirichlet problem for each of the following elliptic equations.

   (a) $\Delta u = e^u$
   (b) $\Delta u = u^2$
   (c) $\Delta u = u^3$
   (d) $\Delta u = u^4$.

16. Consider approximating (3.37) by the forward-backward scheme of Section 2.8. Discuss the advantages and disadvantages of such a technique.

CHAPTER IV

NUMERICAL SOLUTION OF PARABOLIC DIFFERENTIAL EQUATIONS

4.1 Introduction

The prototype parabolic differential equation is the heat equation

$$u_{xx} = u_y,$$

and we will examine it first. Because, physically, the variable y represents <u>time</u> in the problems to be studied, we will set $y = t$ and examine the heat equation in its more customary form

(4.1) $$u_{xx} = u_t.$$

Two kinds of problems are of fundamental interest both mathematically and physically with regard to (4.1). These are the initial value problem and the initial-boundary problem, which are defined as follows. In an initial value problem for (4.1), one is given a function $f(x)$ which is continuous for all values of x and one is asked to find a function $u(x,t)$ which is (a) defined and continuous for $-\infty < x < \infty$, $0 \leq t$; (b) satisfies (4.1) for $-\infty < x < \infty$, $0 < t$; and (c) satisfies the initial condition $u(x,0) = f(x)$ at time $t = 0$ for $-\infty < x < \infty$. In an initial-boundary problem, one is given a constant $a > 0$ and three continuous functions $g_1(t)$, $t \geq 0$; $g_2(t)$, $t \geq 0$; $f(x)$,

INTRODUCTION

$0 \leq x \leq a$, and one is asked to find a function $u(x,t)$ which is:

(a) defined and continuous for $t \geq 0$, $0 \leq x \leq a$; (b) satisfies (4.1) on $0 < x < a$, $t > 0$; and (c) satisfies the initial and boundary conditions

(4.2) $\begin{cases} u(x,0) = f(x), & 0 \leq x \leq a \quad \text{(initial condition)} \\ u(0,t) = g_1(t), & t \geq 0 \\ u(a,t) = g_2(t), & t \geq 0 \end{cases}$ (boundary conditions).

As shown in Figure 4.1, initial value problems are defined on a half-plane, while, as shown in Figure 4.2, initial-boundary problems are defined on a semi-infinite strip.

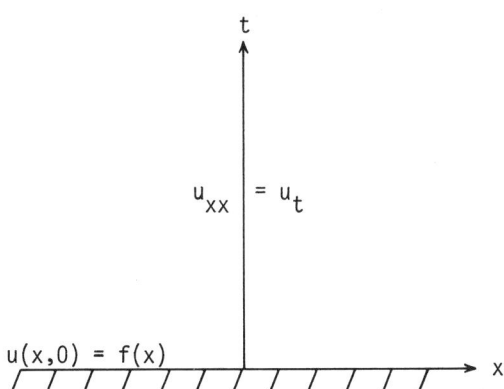

Figure 4.1

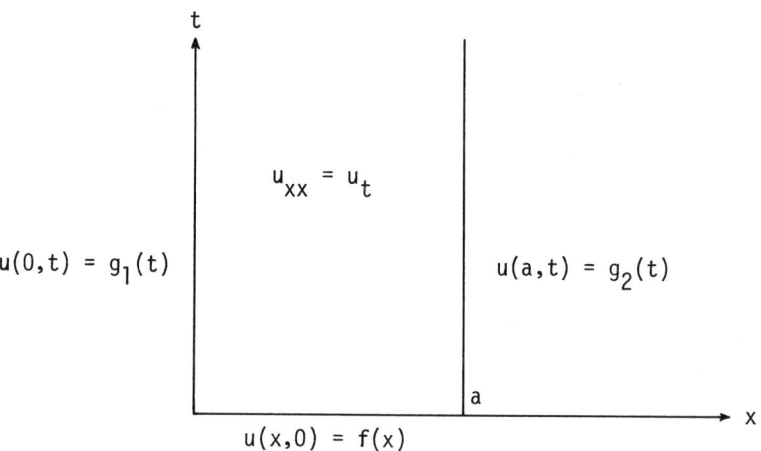

Figure 4.2

The solution of an initial value problem can be given by means of the Fourier integral, while that of an initial-boundary problem can be given in terms of Fourier series (see, e.g., Friedman, Greenspan (2), Petrovsky). However, because the methods for generating such solutions do not extend to nonlinear problems, and because analytical solutions so given are not easily evaluated at particular points of interest, we turn next to numerical methods. For clarity we will concentrate on initial-boundary problems, though most of the ideas extend to the initial value problem also.

4.2 Stability

Because, in the initial-boundary problem, t can vary in the unbounded range $0 \leq t < \infty$, it is necessary to study the possible instability of any numerical method to be devised. It is also of interest to note that one wishes to allow great flexibility in the choice of grid size Δt, for in fast reaction type problems, like those related, for example, to the release of nuclear energy, it is often necessary to choose Δt relatively small in order to generate a physically meaningful numerical solution. On the other hand, for slow reaction type problems, like those related, for example, to radioactive decay, technological and economic limitations require, at present, the choice of a relatively large Δt. To allow for such possibilities, let us begin in a flexible way by choosing grid sizes $\Delta x = h$ and $\Delta t = k$ which are not necessarily equal in magnitude. The grid size h is determined by subdividing $0 \leq x \leq a$ into n equal parts in the usual way. If R is the set of points (x,t) whose coordinates satisfy $0 < x < a$, $t > 0$, and if S is the boundary of R, then a choice of h and k results in a rectangular set of grid points whose interior points R_h are shown as crossed in Figure 4.3 and whose boundary points S_h are shown as circled in Figure 4.3. These grid points lie on the line whose equation is $y = mk$ are called the mth row of grid points.

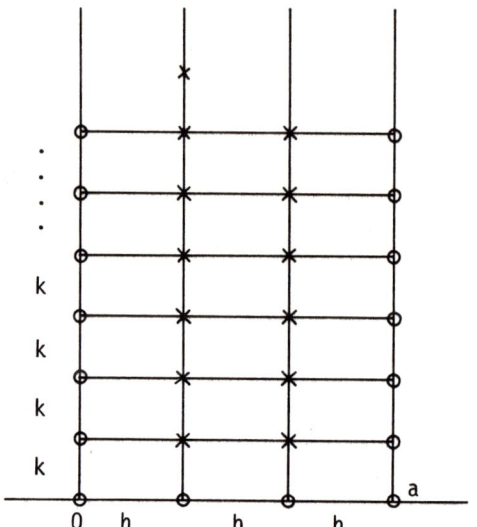

Figure 4.3

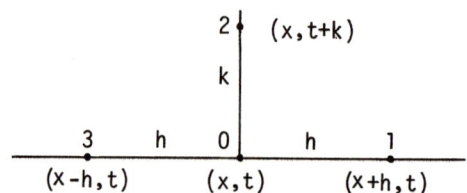

Figure 4.4

STABILITY

For the point arrangement shown in Figure 4.4, consider, from (2.4) and (2.9), the approximations

$$u_{xx}(x,t) = \frac{u(x-h,t) - 2u(x,t) + u(x+h,t)}{h^2}$$

$$u_t(x,t) = \frac{u(x,t+k) - u(x,t)}{k},$$

substitution of which into (4.1) yields the approximation

(4.3) $$\frac{u(x-h,t) - 2u(x,t) + u(x+h,t)}{h^2} = \frac{u(x,t+k) - u(x,t)}{k},$$

or, equivalently,

(4.3') $$u(x,t+k) = u(x,t) + \frac{k}{h^2}[u(x+h,t) - 2u(x,t) + u(x-h,t)].$$

In (4.3'), setting

(4.4) $$\lambda = \frac{k}{h^2}$$

yields finally

(4.5) $$u(x,t+k) = \lambda u(x+h,t) + (1-2\lambda)u(x,t) + \lambda u(x-h,t),$$

which, in the numbering of Figure 4.4 can be written in subscript notation as

(4.6) $$u_2 = \lambda u_1 + (1-2\lambda)u_0 + \lambda u_3.$$

A simple numerical method for approximating a solution of initial-boundary problem (4.1)-(4.2) can be formulated now as follows. Fix h and k and construct R_h and S_h. Apply (4.5), or

(4.6), to approximate u explicitly at each point of the first row of R_h using the known values of u given in (4.2). Using (4.2) and the numerical results generated for row 1, approximate u explicitly at each point of the second row of R_h by means of (4.5), or (4.6). Continue in the indicated fashion to approximate u explicitly at each grid point of row $k+1$, $k = 2, 3, \ldots$, by applying (4.5), or (4.6), and by making use of (4.2) and the numerical approximation generated on row k.

Example

Consider the initial-boundary problem defined by (4.1), $a = 1$, and

(4.7) $\qquad u(0,t) = g_1(t) = 0$

(4.8) $\qquad u(x,0) = f(x) = x, \quad 0 \leq x \leq 1$

(4.9) $\qquad u(1,t) = g_2(t) = 1$,

as shown diagramatically in Figure 4.5. For $h = k = \frac{1}{3}$, construct R_h and S_h, and number the points of R_h as shown in Figure 4.6. Note finally that (4.5) and (4.6) can be written, respectively, as

(4.5') $\qquad u(x, t+\frac{1}{3}) = 3u(x+\frac{1}{3}, t) - 5u(x,t) + 3u(x-\frac{1}{3}, t)$

(4.6') $\qquad u_2 = 3u_1 - 5u_0 + 3u_3$.

STABILITY

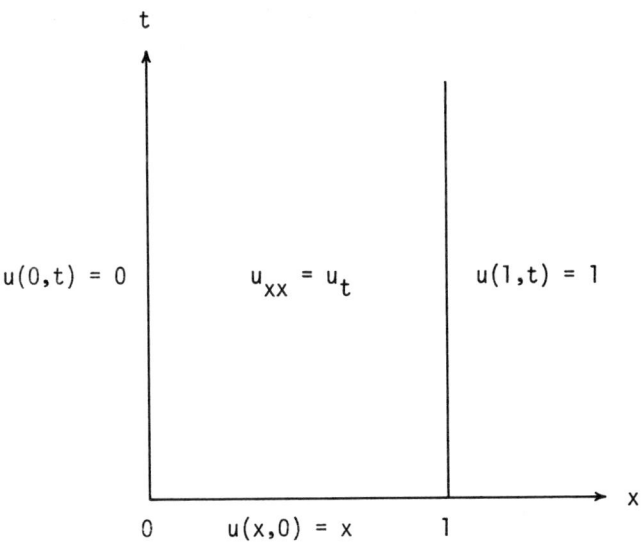

Figure 4.5

Figure 4.6

Rounding all numbers to one decimal place, one has, by applying (4.5'), or (4.6'), successively at the points numbered 1-11 in Figure 4.6, that

$$u_1 = 3u(\tfrac{2}{3},0) - 5u(\tfrac{1}{3},0) + 3u(0,0) = 3(0.7) - 5(0.3) + 3 \cdot 0 = 0.6$$

$$u_2 = 3u(1,0) - 5u(\tfrac{2}{3},0) + 3u(\tfrac{1}{3},0) = 3 \cdot 1 - 5(0.7) + 3(0.3) = 0.4$$

$$u_3 = 3u_2 - 5u_1 + 3u(0,\tfrac{1}{3}) = -1.8$$

$$u_4 = 3u(1,\tfrac{1}{3}) - 5u_2 + 3u_1 = 2.8$$

$$u_5 = 3u_4 - 5u_3 + 3u(0,\tfrac{2}{3}) = 17.4$$

$$u_6 = 3u(1,\tfrac{2}{3}) - 5u_4 + 3u_3 = -16.4$$

$$u_7 = 3u_6 - 5u_5 + 3u(0,1) = -136.2$$

$$u_8 = 3u(1,1) - 5u_6 + 3u_5 = 137.2$$

$$u_9 = 3u_8 - 5u_7 + 3u(0,\tfrac{4}{3}) = 1092.6$$

$$u_{10} = 3u(1,\tfrac{4}{3}) - 5u_8 + 3u_7 = -1091.6$$

$$u_{11} = 3u_{10} - 5u_9 + 3u(0,\tfrac{5}{3}) = -8737.8$$

from which one suspects the development of instability, which must be studied now in some detail.

To begin with, it is important to know that, like harmonic functions, solutions of the heat equation possess the max-min property (Friedman). Any numerical solution which also possesses this property will be called <u>physically reasonable</u>. With this in mind, we define a numerical solution of an initial-boundary problem for (4.5), with continuous boundary data, to be <u>stable</u> if and only if it is physically reasonable. To develop a stability condition for (4.5), consider the following intuitive argument. For a given initial-boundary problem, let $\Delta x = a/2$, $\Delta t = k$, so that the points R_h and S_h are as shown in Figure 4.7. At the point $(\frac{a}{2}, 0)$, set $u = \varepsilon > 0$. At the remaining points of S_h set $u = 0$. Then, application of (4.5) at the points numbered 1, 2, 3,... of R_h yields

(4.10) $\quad u_1 = (1 - 2\lambda)\varepsilon, \; u_2 = (1 - 2\lambda)^2 \varepsilon, \; u_3 = (1 - 2\lambda)^3 \varepsilon, \ldots, u_m = (1 - 2\lambda)^m \varepsilon.$

Now, to have stability, since $0 \le u \le \varepsilon$ on S_h, it follows from the max-min property that for $m = 1, 2, 3, \ldots,$

(4.11) $\quad\quad\quad\quad 0 \le (1 - 2\lambda)^m \varepsilon \le \varepsilon,$

or,

$$0 \le 1 - 2\lambda \le 1.$$

Thus,

$$0 \le \lambda \le \frac{1}{2}.$$

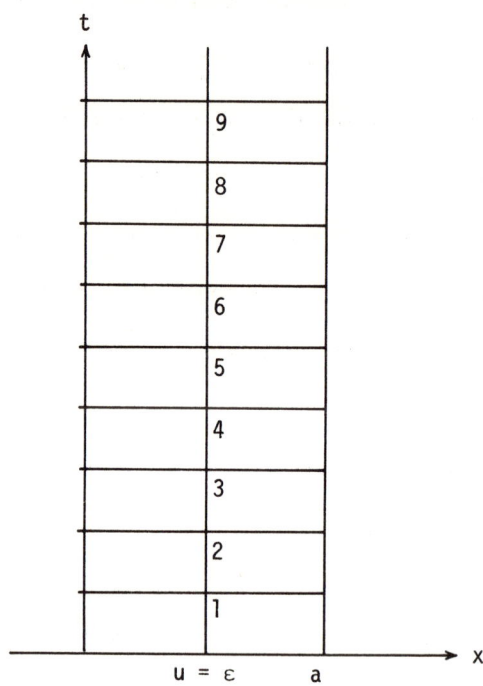

Figure 4.7

But, from (4.4), $\lambda > 0$, so that the stability condition becomes

(4.12) $$0 < \lambda \leq \frac{1}{2}.$$

That (4.12) is in fact the usual stability condition for (4.5) when the given analytical problem has a bounded solution has been established rigorously elsewhere (see, e.g., Collatz (1) Douglas, Forsythe and Wasow).

METHOD I - EXPLICIT

4.3 An Explicit Numerical Method

From the discussion in Section 4.2, one can now formulate as an algorithm the following method for approximating the solution of initial-boundary value problem (4.1)-(4.2).

Method I - Explicit

Step 1 Fix $\Delta x = h$, $\Delta t = k$, so that $\lambda = \frac{k}{h^2} \leq \frac{1}{2}$. Construct R_h and S_h and number the points of R_h.

Step 2 Apply (4.5) to approximate u explicitly with the aid of (4.2) at each point of the first row of R_h.

Step 3 Using (4.2) and the numerical results generated on row k, $k \geq 1$, approximate u explicitly at each point of row $k+1$, $k = 1, 2, \ldots$, by means of (4.5).

Step 4 Terminate the computation when so desired.

Method I has a firm mathematical basis (Forsythe and Wasow), but suffers from a low order of accuracy and relatively severe stability restrictions. It is, however, conceptually and structurally simple. If one wishes to achieve greater accuracy for a given h, or eliminate the stability condition, then one can do either, but at the expense of having to do more work. This leads naturally to the so-called implicit methods, which will be discussed next.

130   PARABOLIC EQUATIONS

4.4   An Implicit Numerical Method

Suppose first that one wishes to construct a method which is stable for <u>all</u> h and k . Such a method may be desirable, for example, if one has to calculate for very long periods of time. If, say, one wishes to have a numerical approximation at t = 100 and one has to choose $h = \frac{1}{100}$, then (4.12) implies that one must choose $\Delta t = k$ to satisfy

$$k \leq \frac{1}{2} \left(\frac{1}{100}\right)^2 = 0.00005 .$$

To generate a numerical solution at t = 100 would therefore require, using Method I, computation on a minimum of two million rows of points of R_h.

Interestingly enough (Ames; Forsythe and Wasow), condition (4.12) can be eliminated simply by replacing the point pattern shown in Figure 4.4 by the one shown in Figure 4.8, or, more precisely, by

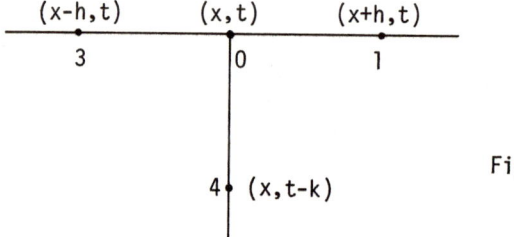

Figure 4.8

METHOD II - IMPLICIT

replacing (4.3) with

(4.13) $$\frac{u(x-h,t) - 2u(x,t) + u(x+h,t)}{h^2} = \frac{u(x,t) - u(x,t-k)}{k}$$

It follows that (4.13) can be written equivalently as

(4.14) $$\lambda u(x-h,t) - (1+2\lambda)u(x,t) + \lambda u(x+h,t) = -u(x,t-k),$$

where λ is defined by (4.4). Using the numbering of Figure 4.8, (4.14) can be written in subscript notation as

(4.15) $$\lambda u_3 - (1+2\lambda)u_0 + \lambda u_1 = -u_4.$$

Though the resulting computation will be stable for <u>all</u> λ, let us show in an illustrative example that the additional work entailed is that of solving a tridiagonal, diagonally dominant system of linear algebraic equations <u>for each row</u> of grid points of R_h.

Example

Consider the initial-boundary problem (4.1), (4.7)-(4.9). For $h = \frac{1}{5}$, $k = 1$, construct R_h and S_h, and number the points of R_h as shown in Figure 4.9. Since $\lambda = 25$, application of (4.14) at the points 1, 2, 3, 4 in Figure 4.9 yields

$$25u(0,1) - 51u_1 + 25u_2 = -u(\tfrac{1}{5},0)$$
$$25u_1 - 51u_2 + 25u_3 = -u(\tfrac{2}{5},0)$$
$$25u_2 - 51u_3 + 25u_4 = -u(\tfrac{3}{5},0)$$
$$25u_3 - 51u_4 + 25u(1,1) = -u(\tfrac{4}{5},0) ,$$

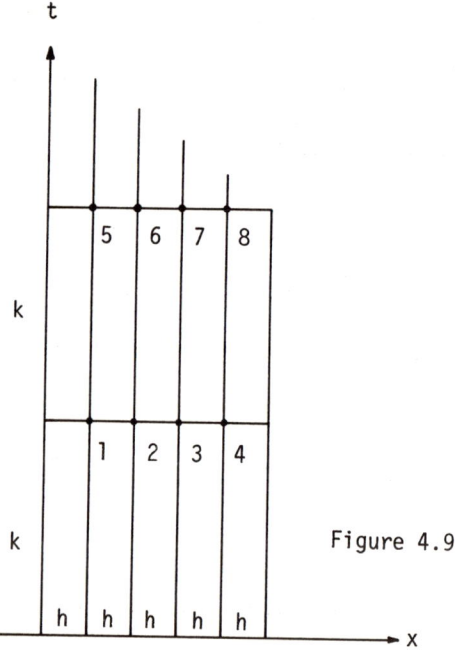

Figure 4.9

or, equivalently,

(4.16)
$$\begin{cases} -51u_1 + 25u_2 & = -\tfrac{1}{5} \\ 25u_1 - 51u_2 + 25u_3 & = -\tfrac{2}{5} \\ 25u_2 - 51u_3 + 25u_4 & = -\tfrac{3}{5} \\ 25u_3 - 51u_4 & = -\tfrac{4}{5} - 25 . \end{cases}$$

METHOD II - IMPLICIT

The solution of system (4.16), found readily by the method of Section 1.4, is given approximately by

$$u_1 = 0.029, \quad u_2 = 0.052, \quad u_3 = 0.059, \quad u_4 = 0.045.$$

On the second row, the system generated by application of (4.14) at the points 5, 6, 7, 8 is

$$25u(0,2) - 51u_5 + 25u_6 = -0.029$$
$$25u_5 - 51u_6 + 25u_7 = -0.052$$
$$25u_6 - 51u_7 + 25u_8 = -0.059$$
$$25u_7 - 51u_8 = -0.045,$$

the solution of which yields the numerical solution on the second row. The method continues in the indicated fashion.

The method illustrated above is called an implicit method because the numerical solution at each point of a given row is generated implicitly in the form of a tridiagonal system, which must then be solved to yield explicitly the approximation at each such point. The method can be described in general by means of the following algorithm.

Method II - Implicit

Step 1 Fix $\Delta x = h$, $\Delta t = k$, construct R_h and S_h, and number the points of R_h.

<u>Step 2</u>   Apply (4.14) at each point of the first row of R_h and, with the aid of (4.2), generate a tridiagonal system of linear algebraic equations.

<u>Step 3</u>   Solve the system generated by Step 2 to yield, explicitly, the numerical solution on the first row of R_h.

<u>Step 4</u>   Apply (4.14) at each point of row m of R_h and, with the aid of (4.2) and the numerical results of row $m-1$, $m = 2,3,\ldots$, generate and solve a tridiagonal system of linear algebraic equations.

<u>Step 5</u>   Terminate the calculations of Step 4 as desired.

4.5   The Crank-Nicolson Method

Since Method II has eliminated the stability restrictions of Method I, the next problem is to improve also on the accuracy of the approximation. For this purpose, note that the use of symmetry in the construction of difference equations can lead to better accuracy in the limited sense that the truncation error is of a higher order of magnitude than when symmetry is not used. Thus, for example, the error in approximation (2.4) is $O(h)$ while that of (2.6), which uses symmetry, is $O(h^2)$. With this notion as an intuitive guide, we can modify Method II to yield greater accuracy in the following simple way. In place of the point pattern shown in Figure 4.8, consider the

CRANK-NICOLSON METHOD

expanded point pattern shown in Figure 4.10. The center of symmetry of the six points shown there is the point $(x, t - \frac{k}{2})$, which is labeled A and is <u>not</u> a grid point. If one wishes, now, to develop formulas symmetrically about A, then note first that

(4.17) $$\left.\frac{\partial u}{\partial t}\right|_A \sim \frac{u(x,t) - u(x,t-k)}{k}$$

does use points symmetrically located about A. Further, since

$$\left.\frac{\partial^2 u}{\partial x^2}\right|_0 \sim \frac{u(x-h,t) - 2u(x,t) + u(x+h,t)}{h^2}$$

and

$$\left.\frac{\partial^2 u}{\partial x^2}\right|_4 \sim \frac{u(x-h,t-k) - 2u(x,t-k) + u(x+h,t-k)}{h^2},$$

it is reasonable to set

$$\left.\frac{\partial^2 u}{\partial x^2}\right|_A = \frac{1}{2}\left[\left.\frac{\partial^2 u}{\partial x^2}\right|_0 + \left.\frac{\partial^2 u}{\partial x^2}\right|_4\right],$$

that is

(4.18) $$\left. u_{xx}\right|_A = \frac{1}{2}\left[\frac{u(x-h,t) - 2u(x,t) + u(x+h,t)}{h^2}\right.$$

$$\left. + \frac{u(x-h,t-k) - 2u(x,t-k) + u(x+h,t-k)}{h^2}\right].$$

Using (4.17) and (4.18) in (4.1) yields

(4.19) $$\frac{1}{2}\left[\frac{u(x-h,t)-2u(x,t)+u(x+h,t)}{h^2}\right.$$

$$\left.+\frac{u(x-h,t-k)-2u(x,t-k)+u(x+h,t-k)}{h^2}\right]$$

$$=\frac{u(x,t)-u(x,t-k)}{k},$$

or, equivalently,

(4.20) $\lambda u(x-h,t) - 2(1+\lambda)u(x,t) + \lambda u(x+h,t) = -\lambda u(x-h,t-k)$

$\quad\quad -2(1-\lambda)u(x,t-k) - \lambda u(x+h,t-k)$.

Using the numbering of Figure 4.10, one can write (4.20) in subscript notation as

(4.21) $\lambda u_3 - 2(1+\lambda)u_0 + \lambda u_1 = -\lambda u_7 - 2(1-\lambda)u_4 - \lambda u_8$.

Formulas (4.20) and (4.21) are called the Crank-Nicolson formulas and, when these are used in Method II in place of (4.14) or (4.15) they do lead to an implicit method which is stable for all λ and which has greater accuracy than Method II. Thus, Method III is given as follows.

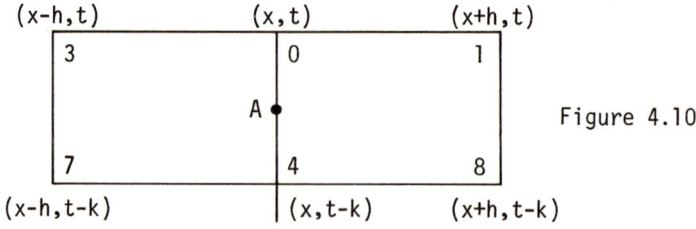

Figure 4.10

CRANK-NICOLSON METHOD

Method III – Crank-Nicolson Implicit

The algorithm is that of Method II with the exception that (4.20) replaces (4.14).

Example

Consider the initial-boundary problem (4.1), (4.7)-(4.9). For $h = \frac{1}{5}$, $k = 1$, construct R_h and S_h and number the points of R_h as shown in Figure 4.9. Since $\lambda = 25$, (4.20) has the form

(4.22)
$$25u(x-h,t) - 52u(x,t) + 25u(x+h,t)$$
$$= -25u(x-h,t-k) + 48u(x,t-k) - 25u(x+h,t-k).$$

If (x,t) is taken, consecutively to be the points numbered 1, 2, 3, 4 in Figure 4.9, then (4.22) yields the tridiagonal linear system

$$25u(0,1) - 52u_1 + 25u_2 = -25u(0,0) + 48u(\tfrac{1}{5},0) - 25u(\tfrac{2}{5},0)$$
$$25u_1 - 52u_2 + 25u_3 = -25u(\tfrac{1}{5},0) + 48u(\tfrac{2}{5},0) - 25u(\tfrac{3}{5},0)$$
$$25u_2 - 52u_3 + 25u_4 = -25u(\tfrac{2}{5},0) + 48u(\tfrac{3}{5},0) - 25u(\tfrac{4}{5},0)$$
$$25u_3 - 52u_4 + 25u(1,1) = -25u(\tfrac{3}{5},0) + 48u(\tfrac{4}{5},0) - 25u(1,0),$$

or, more simply,

(4.23)
$$\begin{cases} -52u_1 + 25u_2 = -\tfrac{2}{5} \\ 25u_1 - 52u_2 + 25u_3 = -\tfrac{4}{5} \\ 25u_2 - 52u_3 + 25u_4 = -\tfrac{6}{5} \\ 25u_3 - 52u_4 = -26\tfrac{3}{5}. \end{cases}$$

One now solves system (4.23) for u_1, u_2, u_3, u_4 and continues to row 2, after which one continues to row 3, and so on.

From the practical point of view, many numerical analysts consider the Crank-Nicolson method to be the most desirable of Methods I-III.

4.6 Mildly Nonlinear Problems

Methods I-III extend in a natural way to mildly nonlinear problems defined by (4.2) and

$$u_{xx} = u_t + f(x,t,u) . \tag{4.24}$$

To assure that solutions of (4.24) have a general max-min property and other important properties, we assume that $|f|$ is bounded and

$$f_u \geq 0 . \tag{4.25}$$

The only modifications necessary in Methods I-III when (4.24) replaces (4.1) are that the difference equations used must be modified appropriately. For Method I, one need only replace (4.5) with

$$u(x,t+k) = \lambda u(x+h,t) + (1-2\lambda)u(x,t) + \lambda u(x-h,t) - kf(x,t,u(x,t)). \tag{4.26}$$

For Method II, one need only replace (4.14) with

$$\lambda u(x-h,t) - (1+2\lambda)u(x,t) + \lambda u(x+h,t) = -u(x,t-k) + kf(x,t-k,u(x,t-k)). \tag{4.27}$$

MILDLY NONLINEAR PROBLEMS

For Method III, in which symmetry was basic, one can use either of the approximations

(4.28) $\quad f \sim f_1 = [f(x,t,u(x,t)) + f(x,t-k,u(x,t-k)]/2$

(4.29) $\quad f \sim f_2 = f(x, \dfrac{t+(t-k)}{2}, \dfrac{u(x,t)+u(x,t-k)}{2})$.

Thus, Method III need be modified only by replacing (4.20) with

(4.30) $\quad \lambda u(x-h,t) - 2(1+\lambda)u(x,t) + \lambda u(x+h,t)$

$$= -\lambda u(x-h,t-k) - 2(1-\lambda)u(x,t-k) - \lambda u(x+h,t-k)$$

$$+ k f_i ,$$

where $i = 1$ or 2 and f_1, f_2 are given by (4.28) and (4.29).

With regard to stability, Methods II and III continue to be stable, in a specialized sense, for all λ, but Method I is stable only for sufficiently small λ (Ames).

From the point of view of actual computation, (4.26) is as simple as (4.5), and (4.27) is as simple as (4.14). Indeed, one can see from (4.27) that the resulting system will be tridiagonal and will have the same coefficient matrix as (4.14). However, (4.30) is not as simple as (4.20), for (4.20) yields a tridiagonal linear system, while (4.30) yields a nonlinear system. However, the generalized Newton's method can be applied to this system because of assumption (4.25). From the point of view of complexity, in general, (4.29) is more cumbersome than (4.28), so that (4.28) is often preferred. This last point can be seen easily by comparing f_1 and f_2 in the simple case $f = u^5$, for, in the numbering of Figure 4.10,

$$f_1 = \frac{1}{2}(u_0^5 + u_4^5),$$

while

$$f_2 = \frac{1}{32}[u_0^5 + 5u_0^4 u_4 + 10u_0^3 u_4^2 + 10u_0^2 u_4^3 + 5u_0 u_4^4 + u_4^5].$$

4.7 A Boundary Value Technique

Methods I-III appear, at present, to be completely adequate for approximating solutions of mildly nonlinear problems. Each is called an initial value, or step-ahead, technique, because results on one row are used to generate results on the next row in a recursive fashion. Inherent in such techniques is an accumulation of the effects of roundoff and truncation, as one proceeds from row to row.

Because the structure and capabilities of future computers is difficult, at present, to predict, and because highly nonlinear problems are not uniformly accessible by any method, we will develop in this final section a boundary value technique for parabolic problems which does not suffer from the row-to-row error accumulation of initial value techniques. The fundamental idea is to determine, a priori, the nature of u as $t \to \infty$, assume these data on a row

BOUNDARY VALUE TECHNIQUE

$t = T$, and then solve the resulting boundary value problem on the truncated region $\bar{R}$: $0 \leq x \leq a$, $0 \leq t \leq T$, shown in Figure 4.11. The method proceeds then as follows.

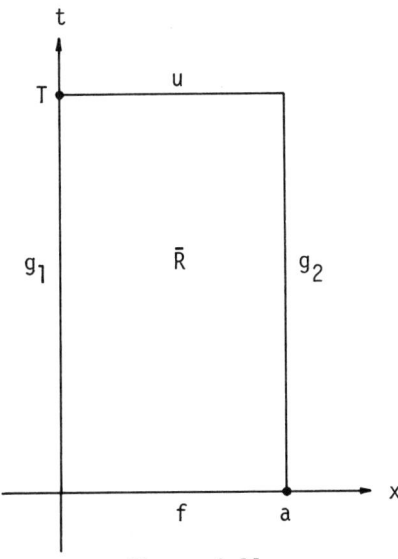

Figure 4.11

First, let us construct a difference analogue of the differential operator

$$u_{xx} - u_t .$$

Let (x,t), $(x+h,t)$, $(x,t+k)$, $(x-h,t)$, $(x,t-k)$ be numbered 0, 1, 2, 3, 4, as shown in Figure 4.12, and set

(4.31) $(u_{xx} - u_t)\big|_0 = \alpha_0 u_0 + \alpha_1 u_1 + \alpha_2 u_2 + \alpha_3 u_3 + \alpha_4 u_4 .$

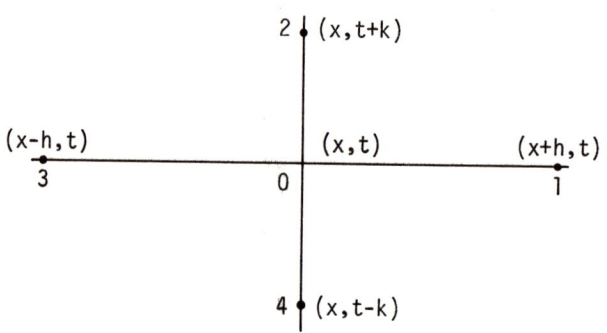

Figure 4.12

Substitution of finite Taylor expansions about (x,t) into (4.31) yields, after recombination of terms,

$$(u_{xx} - u_t)\big|_0 = u_0(\alpha_0 + \alpha_1 + \alpha_2 + \alpha_3 + \alpha_4) + u_x(h\alpha_1 - h\alpha_3)$$
$$+ u_t(k\alpha_2 - k\alpha_4) + u_{xx}(\frac{h^2}{2}\alpha_1 + \frac{h^2}{2}\alpha_3)$$
$$+ u_{tt}(\frac{k^2}{2}\alpha_2 + \frac{k^2}{2}\alpha_4) + O(\alpha_i h^3) + O(\alpha_1 k^3).$$

Setting corresponding coefficients equal implies

$$\alpha_0 + \alpha_1 + \alpha_2 + \alpha_3 + \alpha_4 = 0$$
$$\alpha_1 - \alpha_3 = 0$$
$$\alpha_2 - \alpha_4 = -\frac{1}{k}$$
$$\alpha_1 + \alpha_3 = \frac{2}{h^2}$$
$$\alpha_2 + \alpha_4 = 0,$$

BOUNDARY VALUE TECHNIQUE

the solution of which is

$$\alpha_1 = \alpha_3 = \frac{1}{h^2}, \ \alpha_2 = -\frac{1}{2k}, \ \alpha_4 = \frac{1}{2k}, \ \alpha_0 = -\frac{2}{h^2}.$$

Hence, it is reasonable to take the approximation

(4.32) $\quad (u_{xx} - u_t)\big|_0 \sim -\frac{2}{h^2}u_0 + \frac{1}{h^2}u_1 - \frac{1}{2k}u_2 + \frac{1}{h^2}u_3 + \frac{1}{2k}u_4.$

Note that (4.32) also results easily if one substitutes central differences for u_{xx} and u_t into $u_{xx} - u_t$.

Next, assume that a given parabolic equation of the type

(4.33) $\quad u_{xx} - u_t = f(x,t,u,u_x),$

subject to initial-boundary conditions (4.2), has a solution at $t = \infty$, that is, a steady state solution, which is determined by the ordinary boundary value problem

(4.34) $\quad \dfrac{d^2u}{dx^2} = f(x,u,\dfrac{du}{dx}), \quad 0 \leq x \leq a$

(4.35) $\quad u(0) = \alpha, u(a) = \beta,$

where

(4.36) $\quad \lim_{t \to \infty} g_1(t) = \alpha, \quad \lim_{t \to \infty} g_2(t) = \beta.$

Then the algorithm for the approximate solution of the initial-boundary value problem defined by (4.2) and (4.33) can be given as follows.

Method IV - Boundary Value Technique

Step 1 Divide $0 \leq x \leq a$ into n equal parts, each of length $\Delta x = \frac{a}{n} = h$, by the points $0 = x_0 < x_1 < x_2 < \cdots < x_n = a$. Find either the exact solution, or by the method of Section 2.8, an approximate solution of boundary value problem (4.34)-(4.36). Denote this solution by

(4.37) $$u(x_i, \infty), \quad i = 0, 1, \ldots, n.$$

Step 2 Fix $T > 0$ and define $\bar{S}$ as the rectangle with vertices $(0,0)$, $(a,0)$, (a,T), $(0,T)$ and $\bar{R}$ as its interior. Divide $0 \leq t \leq T$ into m equal parts, each of length $\Delta t = \frac{T}{m} = k > \frac{h^2}{2}$ (this condition will be discussed later), and construct $\bar{R}_h$ and $\bar{S}_h$. Number the points of $\bar{R}_h$ as in Method D for elliptic problems.

Step 3 Define $u(x_i, T)$ by

(4.38) $$u(x_i, T) = u(x_i, \infty), \quad i = 1, 2, \ldots, n-1,$$

so that, from (4.2) and (4.38), u is now defined on all of $\bar{S}_h$.

Step 4 At each point of $\bar{R}_h$, as shown in Figure 4.12, write down, in order, the equation which results by applying

BOUNDARY VALUE TECHNIQUE

(4.39) $\quad -\frac{2}{h^2} u(x,t) + \frac{1}{h^2} u(x+h,t) - \frac{1}{2k} u(x,t+k) + \frac{1}{h^2} u(x-h,t)$

$$+ \frac{1}{2k} u(x,t-k) = f(x,t,u(x,t), \frac{u(x+h,t) - u(x-h,t)}{2h}),$$

inserting the known boundary values whenever possible.

<u>Step 5</u>   Solve the system generated in Step 4, by, say, the generalized Newton's method, to yield the numerical solution on $\bar{R}_h$.

Example 1

For $a = 1$, consider the initial-boundary problem defined by

(4.40) $\quad\quad\quad u_{xx} - u_t = xu_x, \quad\quad 0 \le x \le 1$

(4.41) $\quad\quad\quad u(0,t) = g_1(t) = 0, \quad\quad t \ge 0$

(4.42) $\quad\quad\quad u(x,0) = f(x) = x, \quad\quad 0 \le x \le 1$

(4.43) $\quad\quad\quad u(1,t) = g_2(t) = e^{-t}, \quad t \ge 0.$

The steady state form of (4.40) is

(4.44) $\quad\quad\quad \frac{d^2 u}{dx^2} - x \frac{du}{dx} = 0,$

while α and β, defined in (4.36), are, from (4.41) and (4.43), zero. Thus, the boundary conditions for (4.44) are

(4.45) $\quad\quad\quad u(0) = u(1) = 0.$

Next set $h = \frac{1}{3}$, so that $0 \leq x \leq 1$ is divided into three equal parts by the points $x_0 = 0$, $x_1 = \frac{1}{3}$, $x_2 = \frac{2}{3}$, $x_3 = 1$. Since the analytical solution of (4.44)-(4.45) is $u(x) = 0$, we simply set

(4.46) $\quad u(x_0,\infty) = u(x_1,\infty) = u(x_2,\infty) = u(x_3,\infty) = 0$.

If the analytical solution of (4.44)-(4.45) were not known, then an approximate solution would have been constructed by the method of Section 2.8.

Now, let $T = 2$ and $k = \frac{1}{3}$, so that $\overline{R}_h$ and $\overline{S}_h$ are as shown in Figure 4.13. On $\overline{S}_h$, one has, from (4.38), (4.41)-(4.43) and (4.46),

$$(4.47) \begin{cases} u(0,2) = 0, \ u(0,\tfrac{5}{3}) = 0, \ u(0,\tfrac{4}{3}) = 0, \ u(0,1) = 0, \\ u(0,\tfrac{2}{3}) = 0, \ u(0,\tfrac{1}{3}) = 0 \\ u(0,0) = 0, \ u(\tfrac{1}{3},0) = \tfrac{1}{3}, \ u(\tfrac{2}{3},0) = \tfrac{2}{3}, \ u(1,0) = 1, \\ u(1,\tfrac{1}{3}) = e^{-\tfrac{1}{3}}, \ u(1,\tfrac{2}{3}) = e^{-\tfrac{2}{3}} \\ u(1,1) = e^{-1}, \ u(1,\tfrac{4}{3}) = e^{-\tfrac{4}{3}}, \ u(1,\tfrac{5}{3}) = e^{-\tfrac{5}{3}}, \ u(1,2) = e^{-2}, \\ u(\tfrac{2}{3},2) = 0, \ u(\tfrac{1}{3},2) = 0 \ . \end{cases}$$

BOUNDARY VALUE TECHNIQUE

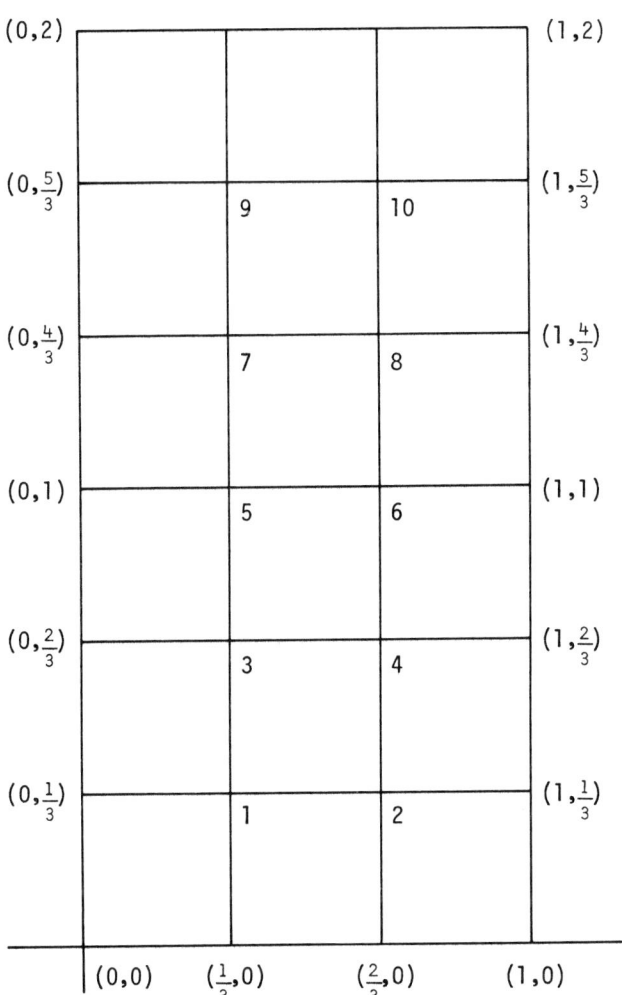

Figure 4.13

PARABOLIC EQUATIONS

Approximation (4.39), as applied to (4.40), takes the form

(4.48) $\quad -18u(x,t) + 9u(x+\frac{1}{3},t) - \frac{3}{2}u(x,t+\frac{1}{3}) + 9u(x-\frac{1}{3},t)$

$\qquad + \frac{3}{2}u(x,t-\frac{1}{3}) = \frac{3x}{2}[u(x+\frac{1}{3},t) - u(x-\frac{1}{3},t)]$,

or, equivalently, in the numbering of Figure 4.12

(4.49) $\quad -2u_0 + (1-\frac{x}{6})u_1 - \frac{1}{6}u_2 + (1+\frac{x}{6})u_3 + \frac{1}{6}u_4 = 0$.

Application of (4.49) at each point of $\bar{R}_h$, with the known values (4.47) inserted, yields the diagonally dominant linear algebraic system:

(4.50)

$$-2u_1 + \frac{17}{18}u_2 - \frac{1}{6}u_3 = -\frac{1}{18}$$

$$-2u_2 - \frac{1}{6}u_4 + \frac{10}{9}u_1 = -\frac{8}{9}e^{-1/3} - \frac{1}{9}$$

$$-2u_3 + \frac{17}{18}u_4 - \frac{1}{6}u_5 + \frac{1}{6}u_1 = 0$$

$$-2u_4 - \frac{1}{6}u_6 + \frac{10}{9}u_3 + \frac{1}{6}u_2 = -\frac{8}{9}e^{-2/3}$$

$$-2u_5 + \frac{17}{18}u_6 - \frac{1}{6}u_7 + \frac{1}{6}u_3 = 0$$

$$-2u_6 - \frac{1}{6}u_8 + \frac{10}{9}u_5 + \frac{1}{6}u_4 = -\frac{8}{9}e^{-1}$$

$$-2u_7 + \frac{17}{18}u_8 - \frac{1}{6}u_9 + \frac{1}{6}u_5 = 0$$

$$-2u_8 - \frac{1}{6}u_{10} + \frac{10}{9}u_7 + \frac{1}{6}u_6 = -\frac{8}{9}e^{-4/3}$$

$$-2u_9 + \frac{17}{18}u_{10} + \frac{1}{6}u_7 = 0$$

$$-2u_{10} + \frac{10}{9}u_9 + \frac{1}{6}u_8 = -\frac{8}{9}e^{-5/3}$$

BOUNDARY VALUE TECHNIQUE

whose solution, when found by the generalized Newton's method, agreed with the analytical solution $u = xe^{-t}$ of (4.40)-(4.43) to at least two, but usually more, decimal places.

Before continuing to the next example, note that in Example 1, the particular choice of h, k and T resulted in diagonal dominance of (4.50). If one must choose T relatively large, then the condition $k > \frac{h^2}{2}$ of Step 2, Method IV, would imply

$$(4.51) \qquad \frac{1}{h^2} > \frac{1}{2k}.$$

But, if (4.33) is linear, (4.39) would always yield a linear algebraic system which is, at least, mildly diagonally dominant, independently of the choice of T. In (4.50), the choice $T = 2$ actually resulted in diagonal dominance.

Example 2

For $a = 1$, consider the initial-boundary problem defined by (4.41)-(4.43) and the non-homogeneous, nonlinear Burger's equation

$$(4.52) \qquad u_{xx} = u_t + uu_x + xe^{-t}(1 - e^{-t}).$$

Proceeding as in Example 1, but with $h = k = \frac{1}{10}$, $T = 10$, and with the difference approximation

$$(4.53) \qquad -2u_0 + u_1 - \frac{1}{20} u_2 + u_3 + \frac{1}{20} u_4 = \frac{1}{20} u_0(u_1 - u_3) + [xe^{-t}(1 - e^{-t})]/100$$

resulted in a nonlinear system of 891 algebraic equations which was solved on the UNIVAC 1108 in 8 seconds by the generalized Newton's method with $\omega = 1.3$ and with a zero initial vector. The numerical solution agreed with $u = xe^{-t}$, the exact solution of the problem, to at least five decimal places and, on the average, to seven.

For the mathematical theory which supports the viability of Method IV, see Carasso and Parter.

Exercises

1. Given the initial-boundary problem for $u_{xx} = u_t$ with $a = 1$, $g_1(t) = 0$, $g_2(t) = 1$, and $f(x) = x^2$, find the numerical solution by Method I on rows 1-10 for each of the following choices.

   (a) $h = 1/4$, $k = 1/10$

   (b) $h = 1/4$, $k = 1/20$

   (c) $h = 1/4$, $k = 1/40$

   (d) $h = 1/4$, $k = 1/80$.

   Which of the above calculations are stable? Which will lead, eventually, to overflow? Which possess the max-min property?

2. For the initial-boundary problem given in Exercise 1, solve on rows 1 and 2 by Method II with $h = 1/5$, $k = 1$.

3. For the initial-boundary problem given in Exercise 1, solve on rows 1 and 2 by Method III with $h = 1/5$, $k = 1$.

4. By the appropriate modification of each of Methods I-III, and by Method IV, find numerical solutions at $t = 3$ for the initial-boundary problem defined by

   $$u_{xx} - u_t = xe^{-t}(t-1), \quad 0 < x < 1, \; t > 0$$

   $$f(x) = 0, \quad 0 \le x \le 1$$

   $$g_1(t) = 0, \quad g_2(t) = te^{-t}, \quad t \ge 0.$$

   Compare your answers with the exact solution $u = xte^{-t}$.

5. By the appropriate modification of each of Methods I-III, and by Method IV, find numerical solutions at $t = 3$ for the initial-boundary problem defined by

$$u_{xx} - u_t = \arctan u, \quad 0 < x < 1, \; t > 0$$

$$f(x) = x, \quad 0 \le x \le 1$$

$$g_1(t) = 0, \quad g_2(t) = e^{-t}, \quad t \ge 0.$$

CHAPTER V

NUMERICAL SOLUTION OF THE WAVE EQUATION

5.1 Introduction

One can study the wave equation profitably either as a second order partial differential equation or as an equivalent system of two first order equations. In this chapter we will study it directly as a second order equation in the same spirit as that of Chapters III and IV. In Chapter VII, in connection with gas dynamical problems, we will study general hyperbolic systems, and that discussion will apply, in particular, to the systems approach to the wave equation.

Since, physically, the variable y will represent time, we will study the wave equation in its more customary form

$$(5.1) \qquad u_{xx} - u_{tt} = 0.$$

Two kinds of problems are of fundamental interest both mathematically and physically with regard to (5.1). These are the Cauchy problem and the initial-boundary problem, which are defined precisely as follows. A Cauchy problem for (5.1) is an initial value problem in which one must find a function $u(x,t)$ which is defined and continuous for $-\infty < x < \infty$, $0 \le t$; which satisfies (5.1) for $-\infty < x < \infty$, $0 < t$; and which satisfies the initial conditions

(5.2) $\quad u(x,0) = f_1(x), \quad -\infty < x < \infty$

(5.3) $\quad u_t(x,0) = f_2(x), \quad -\infty < x < \infty$,

where f_1 and f_2 are given functions of x. An initial-boundary problem is one in which one is given a positive constant a and four continuous functions $g_1(t)$, $t \geq 0$; $g_2(t)$, $t \geq 0$; $f_1(x)$, $0 \leq x \leq a$; $f_2(x)$, $0 < x < a$, and one is asked to find a function $u(x,t)$ which is continuous for $t \geq 0$, $0 \leq x \leq a$; satisfies (5.1) for $0 < x < a$, $t > 0$; and which satisfies the initial and boundary conditions

(5.4) $\quad u(x,0) = f_1(x), \quad 0 \leq x \leq a$ $\Big\}$ initial conditions

(5.5) $\quad u_t(x,0) = f_2(x), \quad 0 < x < a$

(5.6) $\quad u(0,t) = g_1(t), \quad t \geq 0$ $\Big\}$ boundary conditions.

(5.7) $\quad u(a,t) = g_2(t), \quad t \geq 0$

As shown in Figure 5.1, the Cauchy problem is defined on a half-plane, while, as shown in Figure 5.2, initial-boundary problems are defined on a semi-infinite strip.

The solution of the Cauchy-Problem can be given by means of the formula of D'Alembert, while that of an initial-boundary problem can be given in terms of Fourier series (see, e.g., Courant and Hilbert, Greenspan (2), Petrovsky). However, the methods for generating such solutions do not extend to nonlinear problems, and ana-

CAUCHY PROBLEM

lytical solutions so given are usually not evaluated easily at particular points of interest. We will turn then to numerical methods, the development of which will be facilitated by first reviewing the D'Alembert formula.

5.2 The Cauchy Problem

It is well known (see, e.g., Greenspan (2)) that the solution of the Cauchy problem can be given by the D'Alembert formula

(5.8) $$u(x,t) = \frac{1}{2}[f_1(x+t) + f_1(x-t) + \int_{x-t}^{x+t} f_2(r)dr].$$

Before studying (5.8) in detail, let us outline how it can be derived.

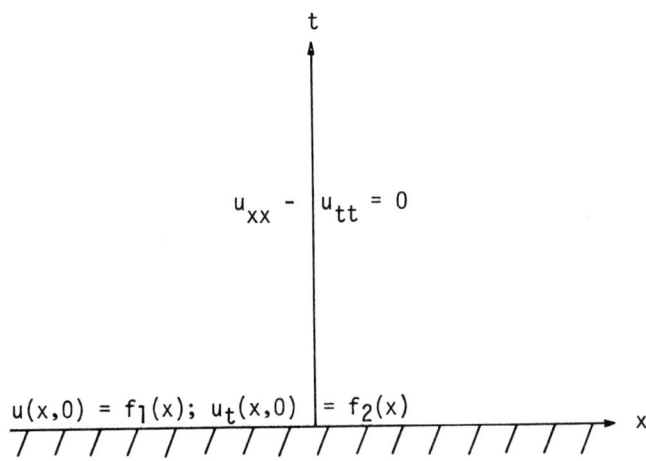

Figure 5.1

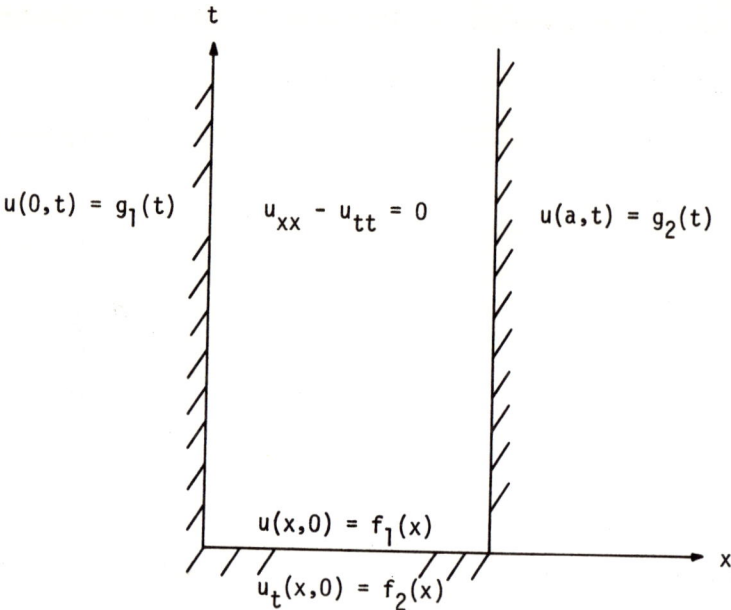

Figure 5.2

Under the change of variables

(5.9) $\quad\quad\quad\quad \xi = x + t, \quad \eta = x - t$,

the wave equation is equivalent to

(5.10) $\quad\quad\quad\quad u_{\xi\eta} = 0$.

Integrating (5.10) yields, first,

$$u_\xi = F_1(\xi),$$

CAUCHY PROBLEM

and, then,

$$u = \int_0^\xi F_1(r)dr + G_2(\eta),$$

where F_1, G_2 are arbitrary differentiable functions. Setting

$$G_1(\xi) = \int_0^\xi F_1(r)dr$$

yields

(5.11) $$u = G_1(\xi) + G_2(\eta).$$

From (5.11), one has

(5.12) $$u_t = \frac{\partial G_1}{\partial \xi}\frac{\partial \xi}{\partial t} + \frac{\partial G_2}{\partial \eta}\frac{\partial \eta}{\partial t} = \frac{\partial G_1(\xi)}{\partial \xi} - \frac{\partial G_2(\eta)}{\partial \eta}.$$

From (5.11) and (5.12), it follows, with the aid of (5.2), (5.3) and (5.9) that

(5.13) $$u(x,0) = G_1(x) + G_2(x) = f_1(x)$$

(5.14) $$u_t(x,0) = G_1'(x) - G_2'(x) = f_2(x).$$

From (5.13), then,

(5.15) $$G_1'(x) + G_2'(x) = f_1'(x),$$

so that, from (5.14) and (5.15),

(5.16) $$G_1'(x) = \frac{1}{2}[f_1'(x) + f_2(x)]; \quad G_2'(x) = \frac{1}{2}[f_1'(x) - f_2(x)].$$

By integration, (5.16) implies

(5.17) $\quad G_1(x) = \frac{1}{2}[f_1(x) + \int_0^x f_2(r)dr]; \; G_2(x) = \frac{1}{2}[f_1(x) - \int_0^x f_2(r)dr].$

Thus, from (5.11) and (5.17),

$$u(x,t) = \frac{1}{2}[f_1(x+t) + \int_0^{x+t} f_2(r)dr] + \frac{1}{2}[f_1(x-t) - \int_0^{x-t} f_2(r)dr],$$

which reduces readily to the D'Alembert formula (5.8).

Example

Find the general solution of the Cauchy problem with $f_1(x) = x^2$, $f_2(x) = e^{-x^2}$.

Solution

By the formula of D'Alembert

(5.18) $\quad u(x,t) = \frac{1}{2}[(x+t)^2 + (x-t)^2 + \int_{x-t}^{x+t} e^{-r^2} dr].$

If one wishes to evaluate the solution of a Cauchy problem at a particular point, then, one can do so either exactly, or, at worst, approximately, if numerical integration is a necessity, from (5.8). Because of this simple state of affairs, we will not pursue the Cauchy problem further, but, instead, will merely make some observations about (5.8) which are fundamental in the study of initial-boundary problems.

CAUCHY PROBLEM

Suppose one is given a Cauchy problem and wishes to know the solution at a point $(\bar{x},\bar{t})$, as shown in Figure 5.3. From (5.8),

(5.19) $\quad u(\bar{x},\bar{t}) = \frac{1}{2}[f_1(\bar{x}+\bar{t}) + f_1(\bar{x}-\bar{t}) + \int_{\bar{x}-\bar{t}}^{\bar{x}+\bar{t}} f_2(r)dr]$.

From (5.19), it follows that $u(\bar{x},\bar{t})$ is determined <u>completely</u> by a knowledge of f_1 and f_2 <u>only</u> between the two points $(\bar{x}-\bar{t},0)$ and $(\bar{x}+\bar{t},0)$ on the X-axis, as shown in Figure 5.3. The interval $\bar{x}-\bar{t} \le x \le \bar{x}+\bar{t}$ is therefore called the interval of dependence for the point $(\bar{x},\bar{t})$. The region interior to the triangle with vertices $(\bar{x},\bar{t})$, $(\bar{x}+\bar{t},0)$, $(\bar{x}-\bar{t},0)$ is called the region of dependence. The line through $(\bar{x},\bar{t})$ and $(\bar{x}+\bar{t},0)$ and the line through $(\bar{x},\bar{t})$ and $(\bar{x}-\bar{t},0)$ are called the <u>characteristics</u> of the wave equation through $(\bar{x},\bar{t})$. The equations of these characteristics are

$$t - \bar{t} = x - \bar{x}, \quad t - \bar{t} = -(x - \bar{x}) \ .$$

Finally, it is important to note that, as a consequence of the above discussion, if one has to solve a Cauchy problem but has been given initial conditions only on $0 \le x \le a$, then one can only find the solution in the region of dependence determined by the point $(\bar{x},\bar{t}) = (\frac{a}{2}, \frac{a}{2})$, that is, in the triangle whose vertices are $(\frac{a}{2}, \frac{a}{2})$, $(0,0)$ and $(a,0)$.

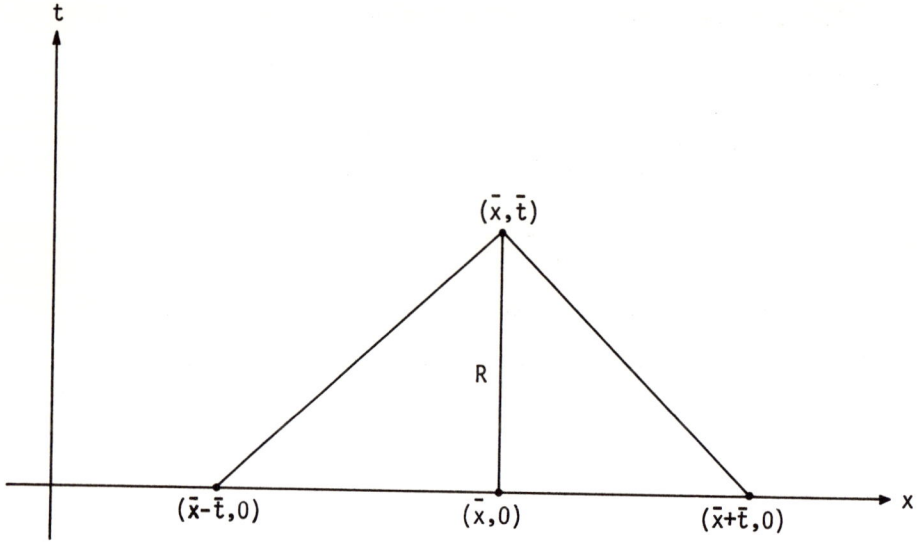

Figure 5.3

5.3 Stability

Let us begin to study initial-boundary problems by developing some intuition with regard to difference approximations for the wave equation. Divide $0 \leq x \leq a$ into n equal parts, each of length $\Delta x = \frac{a}{n} = h$. Let $\Delta t = k$ be arbitrary, at present. If R is the set of all points (x,t) whose coordinates satisfy $0 < x < a$ and $t > 0$, and if S is the boundary of R, then construct R_h and S_h as for the parabolic problem in Section 4.2.

STABILITY

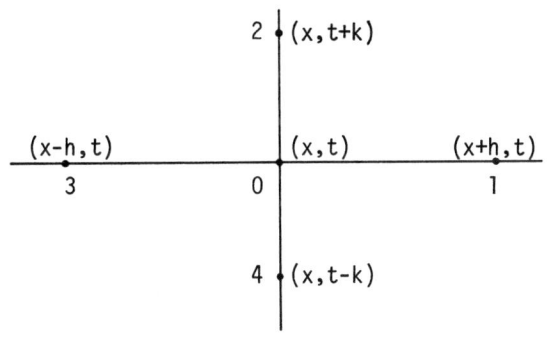

Figure 5.4

By means of (2.9), and for the point arrangement shown in Figure 5.4, consider first the elementary difference approximations

$$u_{xx}(x,t) = [u(x-h,t) - 2u(x,t) + u(x+h,t)]/h^2$$

$$u_{tt}(x,t) = [u(x,t+k) - 2u(x,t) + u(x,t-k)]/k^2,$$

so that the resulting difference approximation for (5.1) is

(5.20) $\quad \dfrac{u(x-h,t) - 2u(x,t) + u(x+h,t)}{h^2} - \dfrac{u(x,t+k) - 2u(x,t) + u(x,t-k)}{k^2} = 0,$

or, equivalently,

(5.21) $\quad u(x,t+k) = 2u(x,t) - u(x,t-k) + \dfrac{k^2}{h^2}[u(x-h,t) - 2u(x,t) + u(x+h,t)].$

Now (5.21) is an explicit formula for $u(x,t+k)$ in terms of values of u on $y = t$ and $y = t - k$, that is, (5.21) is an explicit formula for generating u at any point of a row in terms of values of u on the previous two rows. Thus, it does not appear that (5.21) can be applied on the first row of R_h, but only on the second and higher rows. For this reason, u is approximated on the first row of R_h with the aid of (5.5) by using

$$u_t(x,0) \sim \frac{u(x,k) - u(x,0)}{k} = f_2(x),$$

or, equivalently,

(5.22) $$u(x,k) = u(x,0) + k f_2(x).$$

Consider now a simple illustrative example which will lead directly to the desired stability condition for (5.21). Consider the initial-boundary problem defined by (5.1) and

(5.23) $\quad u(x,0) = x, \quad 0 \le x \le 1$

(5.24) $\quad u_t(x,0) = 1, \quad 0 < x < 1$

(5.25) $\quad u(0,t) = 0, \quad t \ge 0$

(5.26) $\quad u(1,t) = 1, \quad t \ge 0,$

and let us examine the consequences of setting $h = \frac{1}{6}$, $k = \frac{1}{2}$, and of <u>neglecting</u> boundary conditions (5.25)-(5.26). The grid points R_h and S_h are shown in Figure 5.5.

STABILITY

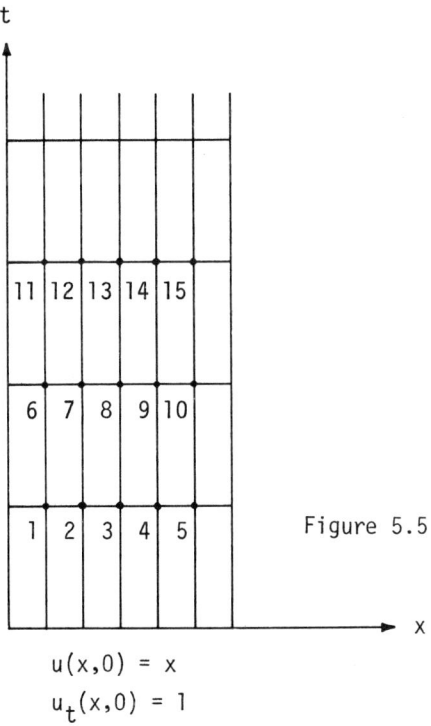

$u(x,0) = x$
$u_t(x,0) = 1$

Figure 5.5

From (5.22) and (5.23), one has the first-row approximation

(5.27) $\quad u_1 = \frac{2}{3}, \quad u_2 = \frac{5}{6}, \quad u_3 = 1, \quad u_4 = \frac{7}{6}, \quad u_5 = \frac{4}{3}$.

Next, u_7, u_8 and u_9 can be generated explicitly from (5.21) to yield

$$u_7 = 2u_2 - u(\tfrac{1}{3},0) + 9[u_1 - 2u_2 + u_3] = \tfrac{4}{3}$$

$$u_8 = 2u_3 - u(\tfrac{1}{2},0) + 9[u_2 - 2u_3 + u_4] = \tfrac{3}{2}$$

$$u_9 = 2u_4 - u(\tfrac{2}{3},0) + 9[u_3 - 2u_4 + u_5] = \tfrac{5}{3}$$

Note that u_6 and u_{10} cannot be so approximated, because we are deliberately neglecting the given boundary data. From u_7, u_8, u_9, then, one can generate, by (5.21)

$$u_{13} = 2u_8 - u_3 + 9[u_7 - 2u_8 + u_9] = 2 .$$

But, now, since we considered only the initial conditions (5.23) and (5.24), u can be determined only in the region of dependence determined by the point $(\frac{1}{2}, \frac{1}{2})$, that is, u can be determined only in the triangle whose vertices are $(0,0)$, $(1,0)$ and $(\frac{1}{2}, \frac{1}{2})$. Since the point $(\frac{1}{2}, \frac{1}{2})$ is numbered 3 in Figure 5.5, it is somewhat unreasonable that any numerical method would yield an approximation for u at the point numbered 13. But this inconsistency is rectified easily by insisting that

(5.28) $$k \leq h ,$$

which will yield approximations only in the domain of dependence. We define, then, a numerical solution generated by (5.20) to be stable if, for all initial data and for zero boundary conditions, it is bounded, and, if the slope of the line through $(x, t+k)$ and $(x-h, t)$ is not greater then the maximum absolute value of the slopes of the characteristics through $(x, t+k)$, that is, $\frac{k}{h} \leq 1$. The latter inequa' is, of course, equivalent to (5.28). Fortunately, (5.28) is also the

METHOD I - EXPLICIT

well-known (Forsythe and Wasow) stability condition for (5.21) when the given initial-boundary problem has a bounded solution.

5.4 An Explicit Method for Initial-Boundary Problems

From the discussion in Section 5.3, one can now formulate in a precise fashion the following method for approximating the solution of initial-boundary problem (5.1), (5.4)-(5.7).

Method I - Explicit

Step 1   Fix $\Delta x = h$ and $\Delta t = k$ so that stability condition (5.28) is satisfied. Construct R_h and S_h and number the points of R_h so that the numbers are increasing from left to right in any row and increasing from bottom to top vertically.

Step 2   Apply (5.22) to approximate u explicitly at each point of the first row of R_h.

Step 3   Apply (5.21) to approximate u explicitly with the aid of (5.4), (5.6), (5.7) and the results of Step 2 at each point of the second row of R_h.

Step 4   Using (5.6), (5.7) and the numerical results for rows k and $k-1$, approximate u explicitly at each point of row $k+1$, $k = 2,3,4,\ldots$, by means of (5.21).

Step 5   Terminate the computation when so desired.

Example

Consider the initial-boundary problem defined by (5.1) and

(5.29) $\qquad u(x,0) = x(1-x) \qquad 0 \le x \le 1$

(5.30) $\qquad u_t(x,0) = 1, \qquad 0 < x < 1$

(5.31) $\qquad u(0,t) = 0, \qquad t \ge 0$

(5.32) $\qquad u(1,t) = 0, \qquad t \ge 0.$

For $h = \frac{1}{4}$ and $k = \frac{1}{8}$, R_h and S_h are shown in Figure 5.6. From (5.22) one has

$$u_1 = u(\tfrac{1}{4}, 0) + \tfrac{1}{8}(1) = \tfrac{5}{16}$$

$$u_2 = u(\tfrac{1}{2}, 0) + \tfrac{1}{8}(1) = \tfrac{3}{8}$$

$$u_3 = u(\tfrac{3}{4}, 0) + \tfrac{1}{8}(1) = \tfrac{5}{16}.$$

Next, since (5.21) has the form

$$u(x,t+k) = 2u(x,t) - u(x,t-k) + \tfrac{1}{4}[u(x-h,t) - 2u(x,t) + u(x+h,t)],$$

or, equivalently,

$$u(x,t+k) = \tfrac{3}{2}u(x,t) - u(x,t-k) + \tfrac{1}{4}[u(x-h,t) + u(x+h,t)],$$

it follows that

METHOD I - EXPLICIT

$$u_4 = \frac{3}{2}u_1 - u(\frac{1}{4}, 0) + \frac{1}{4}[u(0, \frac{1}{8}) + u_2] = \frac{3}{8}$$

$$u_5 = \frac{3}{2}u_2 - u(\frac{1}{2}, 0) + \frac{1}{4}[u_1 + u_3] = \frac{15}{32}$$

$$u_6 = \frac{3}{2}u_3 - u(\frac{3}{4}, 0) + \frac{1}{4}[u_2 + u(1, \frac{1}{8})] = \frac{3}{8}$$

while

$$u_7 = \frac{3}{2}u_4 - u_1 + \frac{1}{4}[u(0, \frac{1}{4}) + u_5] = \frac{47}{128}$$

$$u_8 = \frac{3}{2}u_5 - u_2 + \frac{1}{4}[u_4 + u_6] = \frac{33}{64}$$

$$u_9 = \frac{3}{2}u_6 - u_3 + \frac{1}{4}[u_5 + u(1, \frac{1}{4})] = \frac{47}{128}$$

and so forth.

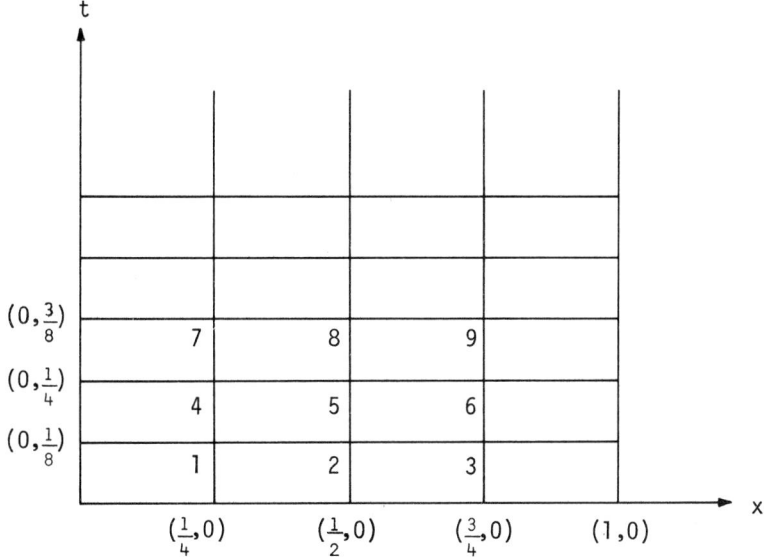

Figure 5.6

5.5 An Implicit Method for Initial-Boundary Problems

Now that a first method, Method I, has been constructed for the solution of initial-boundary problems, one can proceed to try to develop more efficient methods. A simple implicit method, called Method II, which requires the solution of a tridiagonal system for each row of grid points, but which is stable for all choices of h and k, can be constructed as follows (Ames). In the notation of Figure 5.7, use the point (x,t) as a center of symmetry and substitute

$$u_{xx}(x,t) = \frac{1}{2}\{[u(x-h,t+k) - 2u(x,t+k) + u(x+h,t+k)]/[h^2]$$
$$+ [u(x-h,t-k) - 2u(x,t-k) + u(x+h,t-k)]/[h^2]\}$$

$$u_{tt}(x,t) = [u(x,t+k) - 2u(x,t) + u(x,t-k)]/[k^2]$$

into (5.1) to yield

(5.33) $\quad u(x-h,t+k) - 2(1+\frac{h^2}{k^2})u(x,t+k) + u(x+h,t+k) = -u(x-h,t-k)$
$\quad + 2(1+\frac{h^2}{k^2})u(x,t-k) - u(x+h,t-k) - 4\frac{h^2}{k^2}u(x,t).$

In the notation of Figure 5.7, (5.33) can be written more concisely as

(5.34) $\quad u_6 - 2(1+\frac{h^2}{k^2})u_2 + u_5 = -u_7 + 2(1+\frac{h^2}{k^2})u_4 - u_8 - 4\frac{h^2}{k^2}u_0.$

Essentially, the steps of the method are analogous to those of Method I, except that (5.33) replaces (5.21). A formal description proceeds

METHOD II - IMPLICIT

as follows.

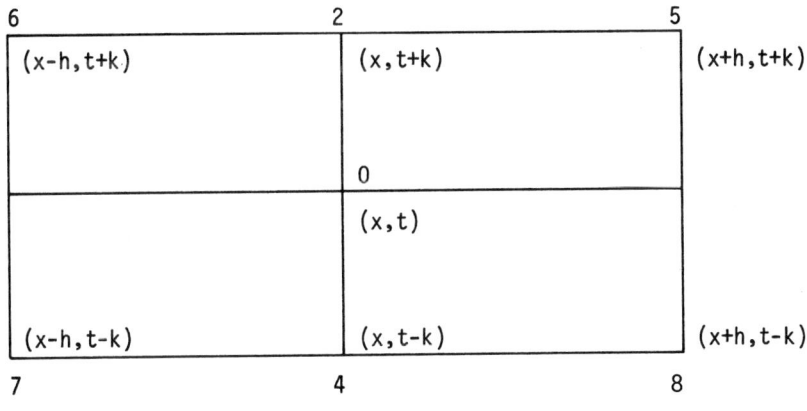

Figure 5.7

Method II - Implicit

Step 1 Execute Steps 1 and 2 of Method I.

Step 2 With the aid of (5.4), (5.6), (5.7) and the results of Step 1, apply (5.33) to generate a tridiagonal system in the unknown numerical approximations on the second row of R_h.

Step 3 Solve the tridiagonal system generated by Step 2 to yield the numerical approximation on row 2 of R_h.

Step 4 Proceed inductively to use the numerical approximations on rows k-1 and k, for k = 2,3,..., by applying (5.33) to generate a tridiagonal system in the unknown numerical approximations on row $k+1$ of R_h, each of which is then determined by solving the resulting algebraic system.

Step 5   Terminate the process when so desired.

Example

Consider the initial-boundary problem defined by (5.1) and (5.29)–(5.32). Set $h = \frac{1}{4}$, $k = \frac{1}{8}$, so that the grid points R_h and S_h are as shown in Figure 5.6. From (5.22) one has, as in the previous example,

$$u_1 = \frac{5}{16}, \quad u_2 = \frac{3}{8}, \quad u_3 = \frac{5}{16}.$$

For the given parameter choices, (5.34) becomes

(5.35) $u_6 - 10u_2 + u_5 = -u_7 + 10u_4 - u_8 - 16u_0.$

Considering first the points numbered 1, 2, 3 in Figure 5.6 to be the point (x,y) in Figure 5.7 yields, in order, by means of (5.35),

$$u(0, \tfrac{1}{4}) - 10u_4 + u_5 = -u(0,0) + 10u(\tfrac{1}{4}, 0) - u(\tfrac{1}{2}, 0) - 16u_1$$

$$u_4 - 10u_5 + u_6 = -u(\tfrac{1}{4}, 0) + 10u(\tfrac{1}{2}, 0) - u(\tfrac{3}{4}, 0) - 16u_2$$

$$u_5 - 10u_6 + u(1, \tfrac{1}{4}) = -u(\tfrac{1}{2}, 0) + 10u(\tfrac{3}{4}, 0) - u(1,0) - 16u_3,$$

or, equivalently,

METHOD III - IMPLICIT

$$-10u_4 + u_5 = -27/8$$
$$u_4 - 10u_5 + u_6 = -31/8$$
$$u_5 - 10u_6 = -27/8 ,$$

the solution of which is

$$u_4 = 43/112, \quad u_5 = 13/28, \quad u_6 = 43/112.$$

One then proceeds to generate the solution on row 3 using (5.35), the given boundary conditions, and the approximations on rows 1 and 2, and so on.

5.6 A Second Implicit Method for Initial Boundary Problems

For a special class of problems, in which the differential equation (5.1) is satisfied on the X-axis and f_2 is given on $0 \leq x \leq a$, one can modify Method II so that one also improves on the accuracy (in the order-of-magnitude sense). The objective is to improve upon (5.22), that is, upon the approximate solution on the very first row of R_h. We will show how to do this by an illustrative example, and then state the method formally.

Example

Consider the initial-boundary problem defined by (5.1) and (5.29)-(5.32). Set $h = \frac{1}{4}$, $k = \frac{1}{8}$, so that the grid points R_h and S_h

are as shown in Figure 5.6. Construct a fictitious set of grid points at $(0, -\frac{1}{8})$, $(\frac{1}{4}, -\frac{1}{8})$, $(\frac{1}{2}, -\frac{1}{8})$, $(\frac{3}{4}, -\frac{1}{8})$, $(1, -\frac{1}{8})$, as shown in Figure 5.8. First, at the three points $(\frac{1}{4}, 0)$, $(\frac{1}{2}, 0)$, $(\frac{3}{4}, 0)$, taken in turn to be the point (x,y) in Figure 5.7, write down (5.33), which, in this case, takes the form (5.35), to yield

$$u(0, \tfrac{1}{8}) - 10u_1 + u_2 = -u(0, -\tfrac{1}{8}) + 10u_{-1} - u_{-2} - 16u(\tfrac{1}{4}, 0)$$

$$u_1 - 10u_2 + u_3 = -u_{-1} + 10u_{-2} - u_{-3} - 16u(\tfrac{1}{2}, 0)$$

$$u_2 - 10u_3 + u(1, \tfrac{1}{8}) = -u_{-2} + 10u_{-3} - u(1, -\tfrac{1}{8}) - 16u(\tfrac{3}{4}, 0).$$

From the given initial and boundary conditions, this system reduces to

(5.36)
$$-10u_1 + u_2 = -u(0, -\tfrac{1}{8}) + 10u_{-1} - u_{-2} - 3$$
$$u_1 - 10u_2 + u_3 = -u_{-1} + 10u_{-2} - u_{-3} - 4$$
$$u_2 - 10u_3 = -u_{-2} + 10u_{-3} - u(1, -\tfrac{1}{8}) - 3.$$

To eliminate $u(0, -\tfrac{1}{8})$, u_{-1}, u_{-2}, u_{-3}, $u(1, -\tfrac{1}{8})$ in this system, utilize central difference approximations for initial condition (5.30) in the form

$$u_t(0,0) = [u(0, \tfrac{1}{8}) - u(0, -\tfrac{1}{8})]/[2k]$$

$$u_t(\tfrac{1}{4}, 0) = [u_1 - u_{-1}]/[2k]$$

$$u_t(\tfrac{1}{2}, 0) = [u_2 - u_{-2}]/[2k]$$

METHOD III - IMPLICIT

$$u_t(\tfrac{3}{4}, 0) = [u_3 - u_{-3}]/[2k]$$

$$u_t(1,0) = [u(1,\tfrac{1}{8}) - u(1,-\tfrac{1}{8})]/[2k],$$

or, equivalently,

(5.37) $\quad u(0,-\tfrac{1}{8}) = -\tfrac{1}{4},\ u_{-1} = u_1 - \tfrac{1}{4},\ u_{-2} = u_2 - \tfrac{1}{4},\ u_{-3} = u_3 - \tfrac{1}{4},\ u(1,-\tfrac{1}{8}) = -\tfrac{1}{4}.$

Substitution of (5.37) into (5.36) yields

$$
\begin{aligned}
-20u_1 + 2u_2 &= -5 \\
2u_1 - 20u_2 + 2u_3 &= -6 \\
2u_2 - 20u_3 &= -5,
\end{aligned}
$$

the solution of which is

(5.38) $\quad u_1 = \tfrac{2}{7},\ u_2 = \tfrac{5}{14},\ u_3 = \tfrac{2}{7}.$

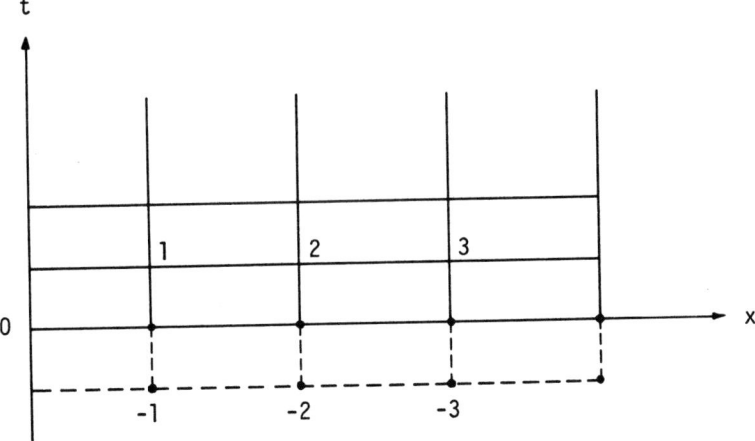

Figure 5.8

One now continues with Step 2 of Method II.

Formally, then, one can describe Method III as follows.

Method III - Implicit

Modify Step 1 of Method II only in the approximation on the first row of R_h. Do this by applying (5.33) at each grid point of the form $(mh,0)$, $m = 1, 2, \ldots, n-1$; by using central differences for $u_t(mh,0)$ to eliminate approximations of u at fictitious grid points that is, at grid points not in R_h or S_h; and by solving the resulting tridiagonal system.

5.7 Mildly Nonlinear Problems

Methods I, II and III extend in a natural way to mildly nonlinear problems defined by (5.4)-(5.7) and

$$(5.39) \qquad u_{xx} - u_{tt} = f(x,t,u).$$

The only modifications necessary occur when (5.39) replaces (5.1), and then one need only replace the difference equations accordingly. In Method I, then, approximate (5.1) by

$$(5.40) \quad \{[u(x-h,t) - 2u(x,t) + u(x+h,t)]/[h^2]\} - \{u(x,t+k) - 2u(x,t) + u(x,t-k)]/[k^2]\} = f(x,t,u(x,t)),$$

MILDLY NONLINEAR PROBLEMS

while, in Methods II and III, approximate (5.1) by any one of

$$\frac{1}{2} \{[u(x-h,t-k) - 2u(x,t-k) + u(x+h,t-k)]/[h^2]$$

$$+ [u(x-h,t+k) - 2u(x,t+k) + u(x+h,t+k)]/[h^2]\}$$

$$- [u(x,t+k) - 2u(x,t) + u(x,t-k)]/[k^2]$$

(5.41)

$$= \begin{cases} f(x,t,u(x,t)) \\ f(x,t,(u(x,t+k) + u(x,t-k))/2) \\ [f(x,t+k,u(x,t+k)) + f(x,t-k,u(x,t-k))]/2 \end{cases}.$$

Computational considerations are analogous to those described in Section 4.6, except that stability conditions vary from problem to problem and no meaningful, comprehensive results are available (Ames). On the computer, one merely experiments with the values of Δx and Δt until one obtains reasonable results.

5.8 A Boundary Value Technique

For the reasons listed in Section 4.7 and 5.7, we will develop, in the spirit of Section 4.7, a boundary value technique for the initial-boundary problem for the wave equation.

First, let us construct a difference analogue of the differential operator

$$u_{xx} - u_{tt}.$$

Since a boundary value technique is to be developed, it will be computationally convenient to construct an analogue which is, at least, mildly diagonally dominant when $h = k$. For this purpose, let (x,t), $(x+h,t)$, $(x, t+k)$, $(x-h, t)$, $(x, t-k)$ be numbered 0, 1, 2, 3, 4, respectively, as shown in Figure 4.12, and set

(5.42) $\quad (u_{xx} - u_{tt})|_0 = \alpha_0 u_0 + \alpha_1 u_1 + \alpha_2 u_2 + \alpha_3 u_3 + \alpha_4 u_4$.

Then, as in the development of (4.32), one finds this time

$$\alpha_0 + \alpha_1 + \alpha_2 + \alpha_3 + \alpha_4 = 0$$

$$\alpha_1 - \alpha_3 = 0, \qquad \alpha_2 - \alpha_4 = 0$$

$$\alpha_1 + \alpha_3 = \frac{2}{h^2}, \qquad \alpha_2 + \alpha_4 = -\frac{2}{k^2},$$

the unique solution of which is

(5.43) $\quad \alpha_1 = \alpha_3 = \frac{1}{h^2}, \quad \alpha_2 = \alpha_4 = -\frac{1}{k^2}, \quad \alpha_0 = -\frac{2}{h^2} + \frac{2}{k^2}$.

But, as observed previously, the diagonal elements of the resulting coefficient matrix will be the numbers α_0, so that, from (5.43), we would like to have

$$\left| \frac{2}{k^2} - \frac{2}{h^2} \right| \geq \max\left(\frac{1}{h^2}, \frac{1}{k^2} \right),$$

BOUNDARY VALUE TECHNIQUE

which is, unfortunately, not valid when $h = k$.

It is natural, then, to ask if an arrangement of points different from that of Figure 4.12 can be used to yield mild diagonal dominance when $h = k$. To study this question, let the points (x,t), $(x+h,t)$, $(x,t+k)$, $(x-h,t)$, $(x,t-k)$, $(x,t-2k)$ be numbered 0, 1, 2, 3, 4, 9, as shown in Figure 5.9. Consider a difference approximation of the wave operator in the form

$$(5.44) \quad (u_{xx} - u_{tt})|_0 = \alpha_0 u_0 + \alpha_1 u_1 + \alpha_2 u_2 + \alpha_3 u_3 + \alpha_4 u_4 + \alpha_9 u_9 .$$

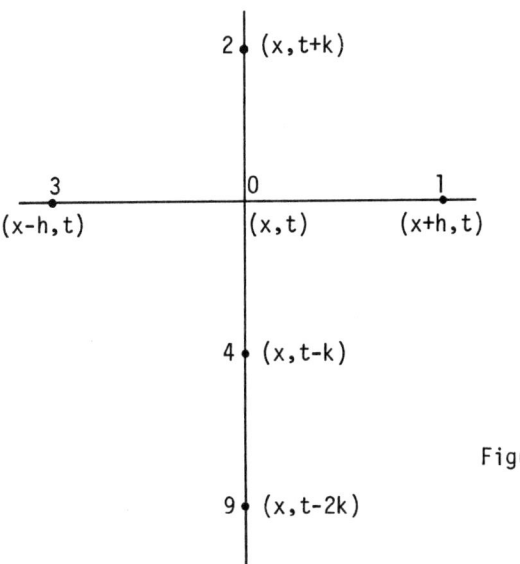

Figure 5.9

Substitution of Taylor expansions about (x,y) for u_1, u_2, u_3, u_4, u_9, and setting corresponding coefficients equal in (5.44) yields the system

$$\alpha_0 + \alpha_1 + \alpha_2 + \alpha_3 + \alpha_4 + \alpha_9 = 0$$

$$\alpha_1 \quad - \alpha_3 \quad\quad\quad = 0$$

$$\alpha_2 \quad\quad - \alpha_4 - 2\alpha_9 = 0$$

$$\alpha_1 \quad + \alpha_3 \quad\quad\quad = 2/h^2$$

$$\alpha_2 \quad\quad + \alpha_4 + 4\alpha_9 = -2/k^2 ,$$

the solution of which, in terms of α_0, is

$$\alpha_1 = \alpha_3 = 1/h^2 \quad,\quad \alpha_4 = -\alpha_0 - \frac{2}{h^2} + \frac{1}{k^2}$$

$$\alpha_2 = -\frac{\alpha_0}{3} - \frac{2}{3h^2} - \frac{1}{3k^2} \quad,\quad \alpha_9 = \frac{\alpha_0}{3} + \frac{2}{3h^2} - \frac{2}{3k^2} .$$

If one chooses

$$\alpha_0 = -\frac{2}{h^2} - \frac{2}{k^2} ,$$

then

$$\alpha_1 = \alpha_3 = 1/h^2 \quad,\quad \alpha_4 = \frac{3}{k^2}$$

$$\alpha_2 = \frac{1}{3k^2} \quad,\quad \alpha_9 = -\frac{4}{3k^2} ,$$

so that

BOUNDARY VALUE TECHNIQUE

(5.45) $(u_{xx} - u_{tt})|_0 \sim -(\frac{2}{h^2} + \frac{2}{k^2})u_0 + \frac{1}{h^2}u_1 + \frac{1}{3k^2}u_2 + \frac{1}{h^2}u_3$
$+ \frac{3}{k^2}u_4 - \frac{4}{3k^2}u_9$.

In order to have mild diagonal dominance, it is sufficient to require

$$\frac{2}{h^2} + \frac{2}{k^2} \geq \frac{3}{k^2},$$

or, equivalently,

(5.46) $$\frac{k^2}{h^2} \geq \frac{1}{2},$$

which is valid when $h = k$.

Note, now, that if one applies the difference analogue

(5.47) $-(\frac{2}{h^2} + \frac{2}{k^2})u_0 + \frac{1}{h^2}u_1 + \frac{1}{3k^2}u_2 + \frac{1}{h^2}u_3 + \frac{3}{k^2}u_4 - \frac{4}{3k^2}u_9 = 0$

for the arrangement of points shown in Figure 5.9, then one has an algebraic equation for each point of R_h except the points of the first row. But this is not unreasonable, because the derivative condition (5.5) has not as yet been considered and this condition can be approximated, with the aid of (2.7), as follows. At each point of S_h of the form $(mh, 0)$, $m = 1, 2, \ldots, n-1$, approximate (5.5) by

(5.48) $[-3u(mh, 0) + 4u(mh, k) - u(mh, 2k)]/(2k) = f_2(mh)$.

With regard to (5.48) note that, since (mh, k) is a point of the first row of R_h, the coefficient of $u(mh, k)$ dominates the other coefficients.

Finally, assume that a given hyperbolic equation of the type

(5.49) $$u_{xx} - u_{tt} = f(x,t,u,u_x,u_t)$$

subject to initial-boundary conditions (5.4)-(5.7) has a solution at $t = \infty$ which is characterized by the boundary value problem

(5.50) $$\frac{d^2u}{dx^2} = f(x,u,\frac{du}{dx}), \qquad 0 \le x \le a$$

(5.51) $$u(0) = \alpha, \qquad u(a) = \beta,$$

where

(5.52) $$\lim_{t \to \infty} g_1(t) = \alpha, \qquad \lim_{t \to \infty} g_2(t) = \beta.$$

The algorithm for the approximate solution of the initial boundary problem defined by (5.1), (5.4)-(5.7) can be given now as follows.

Method IV - Boundary Value Technique

Step 1 Divide $0 \le x \le a$ into n equal parts, each of length $\Delta x = \frac{a}{n} = h$, by the points $0 = x_0 < x_1 < x_2 < \cdots < x_n = a$. Find either the exact solution, or by the method of Section 2.8, an approximate solution on $x_0, x_1, x_2, \ldots, x_{n-1}, x_n$ of boundary value problem (5.50)-(5.52). Denote this solution at $x_0, x_1, \ldots, x_n$ by $u(x_i, \infty)$, $i = 0,1,2,\ldots,n$.

Step 2 Fix $T > 0$ and define $\bar{S}$ as the rectangle with vertices

BOUNDARY VALUE TECHNIQUE

$(0,0)$, $(a,0)$, (a,T), $(0,T)$, and $\bar{R}$ as its interior. Divide $0 \le t \le T$ into m equals parts, each of length $\Delta t = \frac{T}{m} = k$ so that (5.46) is valid and construct $\bar{R}_h$ and $\bar{S}_h$. Number the points of $\bar{R}_h$ as in Method D for elliptic problems.

Step 3 Define $u(x_i, T)$ by

(5.53) $\qquad u(x_i, T) = u(x_i, \infty); \quad i = 1, 2, \ldots, n-1$,

so that, by (5.4), (5.6), (5.7) and (5.53), u is known on all of $\bar{S}_h$.

Step 4 At each point of the first row of $\bar{R}_h$, write down, in order, and in subscript notation, (5.48). On the remainder of $\bar{R}_h$, write down, in order, the difference approximation

(5.54) $\qquad -(\frac{2}{h^2} + \frac{2}{k^2}) u_0 + \frac{1}{h^2} u_1 - \frac{1}{3k^2} u_2 + \frac{1}{h^2} u_3 + \frac{3}{k^2} u_4 - \frac{4}{3k^2} u_9$

$$= f(x, t, u_0, \frac{u_1 - u_3}{2h}, \frac{u_2 - u_4}{2k})$$

of (5.49).

Step 5 Solve the system generated in Step 4 by, say, the generalized Newton's method to yield the numerical solution on $\bar{R}_h$.

Because the mechanics of the numerical method are almost identical to those described in the examples of Section 4.7, we merely remark that several detailed, large scale examples are given in Greenspan (6) and that, as yet, no mathematical theory has been established to support the validity of the method of this section.

5.9   Other Methods

Because of the very extensive scientific interest in waves and wave mechanics, a very large number of related numerical methods have been developed and are worthy of examination. Classical methods which are of more than routine interest, like the method of characteristics, are summarized by Ames. An adaptation of the Runge-Kutta method is given by R. H. Moore and a combination "integral-difference" technique for the reduced wave equation is given by Greenspan and Werner. The method of Garabedian and Lieberstein for detatched waves is also of special interest (see Garabedian).

EXERCISES

Exercises

1. Show that the change of variables $\xi = x+t$, $\eta = x-t$ transforms the wave equation into

$$4u_{\xi\eta} = 0.$$

2. Find the solution of the Cauchy problem for each of the following cases and evaluate $u(1,1)$.

   (a) $f_1 = 1$, $f_2 = -1$

   (b) $f_1 = x$, $f_2 = x^2$

   (c) $f_1 = x^2$, $f_2 = e^{-x^2}$.

3. Find the interval of dependence for each of the following points: $(0,3)$, $(1,3)$, $(-3,3)$, $(7,8)$, $(-7,-1)$, $(-3,6)$.

4. Find the equations of the characteristics through each point of Exercise 3.

5. Given the initial-boundary problem for $u_{xx} - u_{tt} = 0$ with $a = 1$, $g_1(t) = e^{-t}$, $g_2(t) = 2 + e^{1-t}$, $f_1(x) = 2x + e^x$, $f_2(x) = -e^x$, find the numerical solution at $t = 5$ by each of Methods I-III. Compare your results with the exact solution $u = 2x + e^{x-5}$.

6. Given the intial-boundary problem for $u_{xx} - u_{tt} = 0$ with $a = 1$, $g_1(t) = 0$, $g_2(t) = e^{-t}$, $f_1(x) = x$, and $f_2(x) = x^2$, find a numerical solution at $t = 5$.

7. By modifying Methods I-III appropriately, and also by Method IV, find numerical solutions at $t = 4$ of the initial-boundary value problem defined by

$$u_{xx} - u_{tt} = 2(t-x)(t+x+2)u^3$$

$$u(x,0) = \frac{1}{1+x}, \quad u_t(x,0) = -\frac{1}{1+x}$$

$$u(0,t) = \frac{1}{1+t}, \quad u(1,t) = \frac{1}{2(1+t)}.$$

Compare your results with the exact solution

$$u = \frac{1}{(1+x)(1+t)}.$$

CHAPTER VI

APPROXIMATE EXTREMIZATION OF FUNCTIONALS

6.1 Introduction

Problems which are more complex than those which are mildly nonlinear usually require more specialized techniques than those described thus far. Also, the amount of available mathematical support concerning convergence, stability and other fundamental questions usually decreases with the complexity of the problem. Nevertheless, such problems still demand attention and it is to these that the next three chapters are directed. Boundary value problems will be studied first.

6.2 Extremization of Functionals

Historically, one of the oldest mathematical disciplines to be intimately involved with applied problems is the calculus of variations. In developing numerical methods for nonlinear problems in which the defining equations may be more than mildly nonlinear, we shall examine first the classical variational problems in one and two dimensions.

The fundamental problem in the calculus of variations may be formulated as follows. For a, b, α and β real numbers, with a < b, and for given $F(x,y,p)$ which has continuous first order partial derivatives, find a function $y(x)$ which is defined and has continuous first order derivatives for $a \leq x \leq b$, which satisfies the boundary conditions

(6.1) $$y(a) = \alpha, \quad y(b) = \beta,$$

and which minimizes the integral

(6.2) $$J = \int_a^b F(x,y,y')\,dx.$$

Though one can seek, also, to maximize (6.2), for clarity we shall concentrate on the minimization.

Geometrically, as shown in Figure 6.1, the fundamental problem in the calculus of variations requires that, out of <u>all</u> the continuously differentiable functions defined on $a \leq x \leq b$ whose graphs pass through the two points (a,α) and (b,β), one must find that one which minimizes the given integral (6.2).

Because the value of (6.2) depends on a function, and not just on a real number, the integral (6.2) is in reality a function of a function, and therefore is called a <u>functional</u>. Indeed, (6.2) is the prototype functional of the mathematical discipline called <u>functional analysis</u>.

EXAMPLE   187

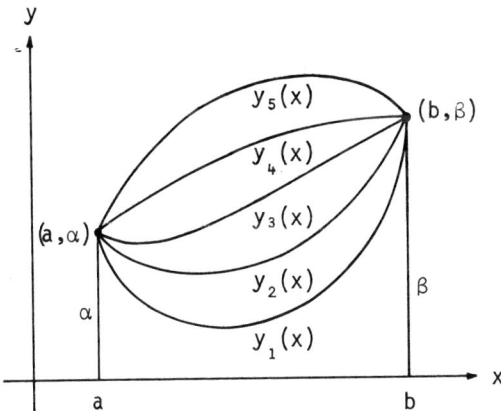

Figure 6.1

Example

Suppose one is given $a = 0$, $b = 1$, $\alpha = 1$, $\beta = 0$, and

(6.3) $$J = \int_0^1 (x^2 y - y') \, dx \ .$$

Then examples of continuously differentiable functions which satisfy the boundary conditions

(6.4) $$y(0) = 1, \quad y(1) = 0$$

are

$$y = 1 - x$$
$$y = 1 - x^3$$
$$y = x^2 - 2x + 1$$

which, when inserted into functional (6.3) yield

$$J(1 - x) = \int_0^1 [x^2(1 - x) - (-1)]dx = \frac{13}{12}$$

$$J(1 - x^3) = \int_0^1 [x^2(1 - x^3) - (-3x^2)]dx = \frac{7}{6}$$

$$J(x^2 - 2x + 1) = \int_0^1 [x^2(x^2 - 2x + 1) - (2x - 2)]dx = \frac{31}{30}.$$

The problem of finding that $y(x)$ which is continuously differentiable on $0 \leq x \leq 1$, which satisfies (6.4), and which minimizes (6.3) is a fundamental type problem of the calculus of variations.

Analytically, the fundamental problem in the calculus of variations is, in general, exceptionally difficult to solve. As in the elementary calculus, where one attempts to find a minimum of a function

(6.5) $$y = f(x)$$

by solving the equation

(6.6) $$f'(x) = 0,$$

so in the calculus of variations one can attempt to find the minimum

EULER'S EQUATION

of a functional

(6.7) $$J = \int_a^b F(x,y,y')\,dx$$

by solving the equation

(6.8) $$\frac{\partial F}{\partial y} - \frac{d}{dx}\frac{\partial F}{\partial y'} = 0,$$

which results by setting what is known as the Frechet derivative of functional (6.7) equal to zero. Equation (6.8) is called the Euler differential equation, and a great portion of the calculus of variations is devoted to the study of the problem defined by (6.1) and (6.8) rather than to the problem defined by (6.1) and (6.2).

Example

The Euler equation of the functional

$$\int_0^1 [xy^3 - (y')^2 + 3xyy']\,dx$$

is

$$(3xy^2 + 3xy') - \frac{d}{dx}(-2y' + 3xy) = 0,$$

or, equivalently,

$$2y'' - 3y + 3xy^2 = 0.$$

Euler differential equation (6.8) is, in general, a nonlinear, second order, ordinary differential equation, and, although such equations are, in general, very difficult to solve, still they seem to be more viable _analytically_ than the functionals from which they are derived. Nevertheless, _numerically_ it is so often easier to approximate a solution of the original variational problem than to solve the problem defined by (6.1) and (6.8), that the approach here will be to examine applied problems which are usually stated in terms of (6.1) and (6.8) by returning to their primitive, variational formulation. Such an approach is also motivated by the observations that for most applied problems one cannot solve the associated Euler differential equation analytically and that, just as a solution of (6.6) yields only an extremal of (6.5), so a solution of (6.8) need not necessarily yield a minimum of (6.7).

6.3 A Numerical Method

For the fundamental problem of the calculus of variations, that is, the boundary value problem defined by (6.1) and (6.2), divide the interval $a \leq x \leq b$ into n equal parts, each of length $h = \frac{b-a}{n}$, by the points $a = x_0 < x_1 < x_2 < \cdots < x_{n-1} < x_n = b$ (see Figure 6.2). Thus, $h = x_i - x_{i-1}$, for $i = 1, 2, \ldots, n$. Let $y_i = y(x_i)$, for $i = 0, 1, 2, \ldots, n-1, n$. Then approximate the functional

(6.9) $$J = \int_a^b F(x, y, y') \, dx$$

by the function

(6.10) $$J_n = h \sum_{i=1}^{n} F(x_{i-1}, y_{i-1}, \frac{y_i - y_{i-1}}{h}).$$

Since, by (6.1), $y_0 = \alpha$ and $y_n = \beta$, it follows that J_n is a function only of $y_1, y_2, \ldots, y_{n-1}$. To find an extremal of J_n, then, consider the system of equations

(6.11) $$\frac{\partial J_n}{\partial y_i} = 0, \quad i = 1, 2, \ldots, n-1.$$

A solution of (6.11) will constitute an approximation at $x_1, x_2, \ldots, x_{n-1}$ of a function $y(x)$ which is a solution of the fundamental problem of the calculus of variations.

The function (6.10) is obtained by a simple rectangular integration approximation of (6.9) in which derivatives are replaced by forward differences. For sufficient criteria for convergence, see, e.g., Greenspan (8). An example will be given in the next section.

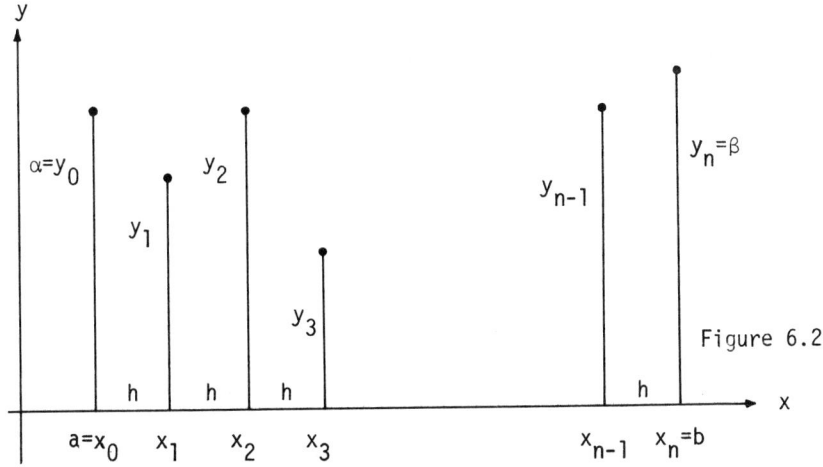

Figure 6.2

6.4 Geodesics

If one wishes to consider motion on the earth or, more generally, in any curved space, then the "shortest" paths between two points is no longer a straight line. The shortest paths between two points in space are called geodesics, and in this section we will illustrate the method of Section 6.3 by applying it to a geodesic problem whose solution is well known.

In ξ, η, ν space, let S be the unit sphere, whose equation is $\xi^2 + \eta^2 + \nu^2 = 1$. Consider the problem of finding the shortest path on S between $(1,0,0)$ and $(\frac{\sqrt{2}}{2}, 0, \frac{\sqrt{2}}{2})$. For this purpose, let S be parametrized by

(6.12)
$$\xi = \cos x \cos y$$
$$\eta = \cos x \sin y$$
$$\nu = \sin x,$$

where x represents latitude and y represents longitude, as shown in Figure 6.3. Then S can be given vectorially by

(6.13) $$\vec{v} = (\cos x \cos y, \cos x \sin y, \sin x)$$

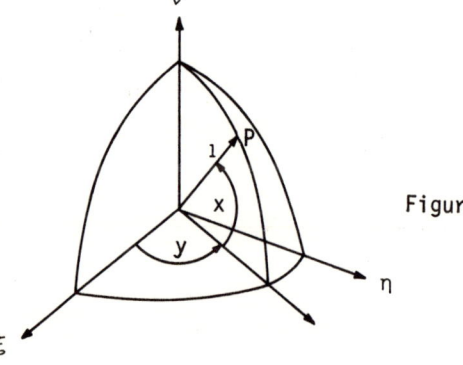

Figure 6.3

GEODESICS

Let the dot products E, F and G be defined by

$$E \equiv \frac{\partial \vec{v}}{\partial x} \cdot \frac{\partial \vec{v}}{\partial x} = 1$$

$$F \equiv \frac{\partial \vec{v}}{\partial x} \cdot \frac{\partial \vec{v}}{\partial y} = 0$$

$$G \equiv \frac{\partial \vec{v}}{\partial y} \cdot \frac{\partial \vec{v}}{\partial y} = \cos^2 x .$$

Since (Struik) geodesics are extrema of the functional

$$\int_{x_1}^{x_2} [E + 2F \frac{dy}{dx} + G(\frac{dy}{dx})^2]^{1/2} dx, \quad y = y(x) ,$$

it follows that we wish to extremize the functional

(6.14) $$J = \int_0^{\pi/4} [1 + (\cos^2 x)(\frac{dy}{dx})^2]^{1/2} dx$$

subject to the boundary conditions

(6.15) $$y(0) = y(\frac{\pi}{4}) = 0 .$$

Setting $h = \pi/16$ implies that $x_0 = 0$, $x_1 = \pi/16$, $x_2 = \pi/8$, $x_3 = 3\pi/16$, $x_4 = \pi/4$ and $y(x_i) = y_i$, $y_0 = y_4 = 0$. Functional (6.14) is then approximated by

(6.16) $$J_4 = \frac{\pi}{16} \sum_{i=1}^{4} \left\{ \left[1 + \left(\cos^2 x_{i-1}\right)\left(\frac{y_i - y_{i-1}}{\pi/16}\right)^2\right]^{1/2} \right\},$$

and system (6.11) takes the form of the following three equations in y_1, y_2 and y_3:

$$\frac{y_1}{\sqrt{\left[1+\left(\frac{y_1}{\pi/16}\right)^2\right]}} - \frac{\left(\cos^2\frac{\pi}{16}\right)(y_2-y_1)}{\sqrt{\left[1+\left(\cos^2\frac{\pi}{16}\right)\left(\frac{y_2-y_1}{\pi/16}\right)^2\right]}} = 0$$

$$\frac{\left(\cos^2\frac{\pi}{16}\right)(y_2-y_1)}{\sqrt{\left[1+\left(\cos^2\frac{\pi}{16}\right)\left(\frac{y_2-y_1}{\pi/16}\right)^2\right]}} - \frac{\left(\cos^2\frac{\pi}{8}\right)(y_3-y_2)}{\sqrt{\left[1+\left(\cos^2\frac{\pi}{8}\right)\left(\frac{y_3-y_2}{\pi/16}\right)^2\right]}} = 0$$

$$\frac{\left(\cos^2\frac{\pi}{8}\right)(y_3-y_2)}{\sqrt{\left[1+\left(\cos^2\frac{\pi}{8}\right)\left(\frac{y_3-y_2}{\pi/16}\right)^2\right]}} - \frac{\left(\cos^2\frac{3\pi}{16}\right)y_3}{\sqrt{\left[1+\left(\cos^2\frac{3\pi}{16}\right)\left(\frac{y_3}{\pi/16}\right)^2\right]}} = 0.$$

The generalized Newton's method with initial guess $y_1 = \frac{1}{2}$, $y_2 = 1$, $y_3 = \frac{1}{2}$ and with $\omega = 1.8$ was applied to solve the above nonlinear algebraic system. On the CDC 3600 the number of iterations was 95 and the running time was 8 sec. The answers were $y_1 = y_2 = y_3 = 0 \cdot 10^{-12}$, so that, from (6.12), one has

$$(\xi_1, \eta_1, \nu_1) = (\cos x_1 \cos y_1, \cos x_1 \sin y_1, \sin x_1)$$
$$= (\cos \frac{\pi}{16}, 0, \sin \frac{\pi}{16})$$

$$(\xi_2, \eta_2, \nu_2) = (\cos x_2 \cos y_2, \cos x_2 \sin y_2, \sin x_2)$$
$$= (\cos \frac{\pi}{8}, 0, \sin \frac{\pi}{8})$$

$$(\xi_3, \eta_3, \nu_3) = (\cos x_3 \cos y_3, \cos x_3 \sin y_3, \sin x_3)$$
$$= (\cos \frac{3\pi}{16}, 0, \sin \frac{3\pi}{16}).$$

Since the shortest path sought is completely determined analytically by $y = 0$, $0 \leq x \leq \frac{\pi}{4}$, the unusual result follows that the above three points actually lie on the resulting geodesic.

For details of examples in which system (6.11) consists of as many as 1500 equations, see Greenspan (8).

6.5 Free Boundary Value Problems

All the ideas and theory presented thus far extend in a natural way to integration formulae other than that incorporated in (6.10) and also to free boundary value problems. We shall illustrate both of these possibilities by means of a single control theory problem.

Consider the problem of minimizing the functional

(6.17) $$J = \int_0^1 [y^2 + (y')^2]dx$$

subject to the boundary conditions

(6.18) $$y(0) = 1, \quad y'(1) = 0 .$$

Setting $h_i = 0.25$ implies $x_0 = 0$, $x_1 = 0.25$, $x_2 = 0.5$, $x_3 = 0.75$, $x_4 = 1$. Functional (6.17) then can be approximated by the trapezoidal integration formula

$$J \sim \frac{1}{8}[y_0^2 + (y_0')^2 + 2y_1^2 + 2(y_1')^2 + 2y_2^2 + 2(y_2')^2 + 2y_3^2 + 2(y_3')^2 + y_4^2 + (y_4')^2] ,$$

which, after the insertion of (6.18), reduces to

(6.19) $\quad J \sim \frac{1}{8}[1 + (y_0')^2 + 2y_1^2 + 2(y_1')^2 + 2y_2^2 + 2(y_2')^2 + 2y_3^2 + 2(y_3')^2 + y_4^2]$.

Next, inserting into (6.19) a forward difference approximation for y_0' and central difference approximations for y_1', y_2' and y_3' yields

$$J \sim J_4 = \frac{1}{8}\left[1 + \left(\frac{y_1 - 1}{0.25}\right)^2 + 2y_1^2 + 2\left(\frac{y_2 - 1}{0.5}\right)^2 + 2y_2^2 + 2\left(\frac{y_3 - y_1}{0.5}\right)^2 \right.$$
$$\left. + 2y_3^2 + 2\left(\frac{y_4 - y_2}{0.5}\right)^2 + y_4^2\right].$$

In order to minimize J_4, the equations:

$$\frac{\partial J_4}{\partial y_i} = 0, \quad i = 1, 2, 3, 4,$$

which, in this case, are equivalent to

$$13y_1 - 4y_3 = 8$$
$$9y_2 - 4y_4 = 4$$
$$4y_1 - 5y_3 = 0$$
$$8y_2 - 9y_4 = 0,$$

yield $y_1 = \frac{40}{49} \sim 0.816$, $y_2 = \frac{36}{49} \sim 0.735$, $y_3 = \frac{32}{49} \sim 0.653$, $y_4 = \frac{32}{49} \sim 0.653$, which compares favorably with the exact solution (0.839, 0.731, 0.668, 0.648), determined from

$$y(x) = [\cosh(1-x)]/[\cosh 1].$$

6.6 Variational Problems and Partial Differential Equations

As regards partial differential equations, the fundamental problem of the calculus of variations can be formulated as follows. Let $F(x,y,u,p,q)$ have continuous first order partial derivatives. Let R be a simply connected, bounded region whose boundary S is piecewise regular, and let $f(x,y)$ be defined and continuous on S. Then one must find a function $u(x,y)$ which has continuous first order partial derivatives on $R + S$, which satisfies

(6.20) $$u \equiv f \text{ on } S$$

and which minimizes

(6.21) $$J = \iint_{R+S} F(x,y,u,u_x,u_y)\, dA .$$

The Euler differential equation of (6.21) is

(6.22) $$F_u - \frac{\partial}{\partial x} F_{u_x} - \frac{\partial}{\partial y} F_{u_y} = 0 .$$

Example

The Euler equation of the functional

$$J = \iint_{R+S} [u_x^2 + u_y^2 + 2e^u - 2]\, dA$$

is

$$u_{xx} + u_{yy} = e^u .$$

We shall study next Dirichlet problems defined by (6.20) and (6.22) by applying to (6.20) and (6.21) a direct generalization of the method developed in Section 6.3. But note that since the methods already developed for mildly nonlinear equations (3.53) and (3.54) are relatively more efficient than the method to be developed in the next section, attention will be directed to (3.55). The reader interested in a variational formulation of equations like (3.53) and (3.54) need note only that the Euler equation of

$$J = \iint_{R+S} [u_x^2 + u_y^2 + 2 \int_0^u G(t)dt]\, dA$$

is

$$\Delta u = G(u).$$

6.7 The Plateau Problem

Let R be a simple connected, bounded region whose boundary S is piecewise regular. Let $f(x,y)$ be defined and continuous on S. Then the Dirichlet problem of finding $u(x,y)$ which is continuous on $R + S$, which satisfies

(6.23) $\qquad\qquad u = f, \text{ on } S$

and which on R satisfies the nonlinear elliptic partial differential equation

THE PLATEAU PROBLEM

(6.24) $\quad (1 + u_y^2)u_{xx} - 2u_x u_y u_{xy} + (1 + u_x^2)u_{yy} = 0$

will be called the Plateau problem.

The Plateau problem is intimately related with the physical problem of soap films, that is, of determining the shape of a soap film which results from having immersed a closed, three dimensional wire into a soap solution. Indeed, (6.24) is the Euler equation of the integral

(6.25) $\quad J = \iint\limits_{R+S} \sqrt{1 + u_x^2 + u_y^2} \, dA ,$

which defines the surface area of the resulting film. In elasticity problems, like those for soap films, one wishes to minimize (6.25).

Rather than discuss the extensive details of an abstract generalization of the method of Section 6.3 to general problems in partial differential equations, we will show simply how to treat these by means of the following illustrative Plateau problem. More complex problems can be treated similarly.

Example

Let S be the square whose vertices are (0,0), (1,0), (1,1), (0,1), and whose interior is denoted by R. On S, define f(x,y) by

(6.26) $\quad f(x,y) = x - 3y$

and consider the associated Plateau problem.

The numerical approach will center about minimizing functional (6.25) subject to the boundary condition (6.26). For this purpose, take $h = \frac{1}{3}$ and construct R_h and S_h. It will be convenient, in the present development, to adjoin to S_h the vertices $(0,0)$, $(1,0)$, $(1,1)$, and $(0,1)$, and this will be done. Number the points of R_h with 1, 2, 3, 4 and those of S_h with 5, 6, 7,...,16, as shown in Figure 6.4. Next, triangulate, in any fashion, each subsquare shown in Figure 6.4, so that $R + S$ is thereby divided into 18 mutually disjoint, subtriangular regions, one possible arrangement of which is shown in Figure 6.5. Notice that the process of triangularization introduces <u>no</u> new grid points. Number these triangular regions R_1, R_2,...,R_{18}, in any order, and let the boundary of each R_i be denoted by S_i, $i = 1, 2, ..., 18$.

Now, note that (6.25) can be rewritten as

$$J = \iint_{R_1+S_1} \sqrt{1+u_x^2+u_y^2}\, dA + \iint_{R_2+S_2} \sqrt{1+u_x^2+u_y^2}\, dA + \cdots$$

$$+ \iint_{R_{18}+S_{18}} \sqrt{1+u_x^2+u_y^2}\, dA ,$$

and consider first

$$I_1 = \iint_{R_1+S_1} \sqrt{1+u_x^2+u_y^2}\, dA .$$

THE PLATEAU PROBLEM

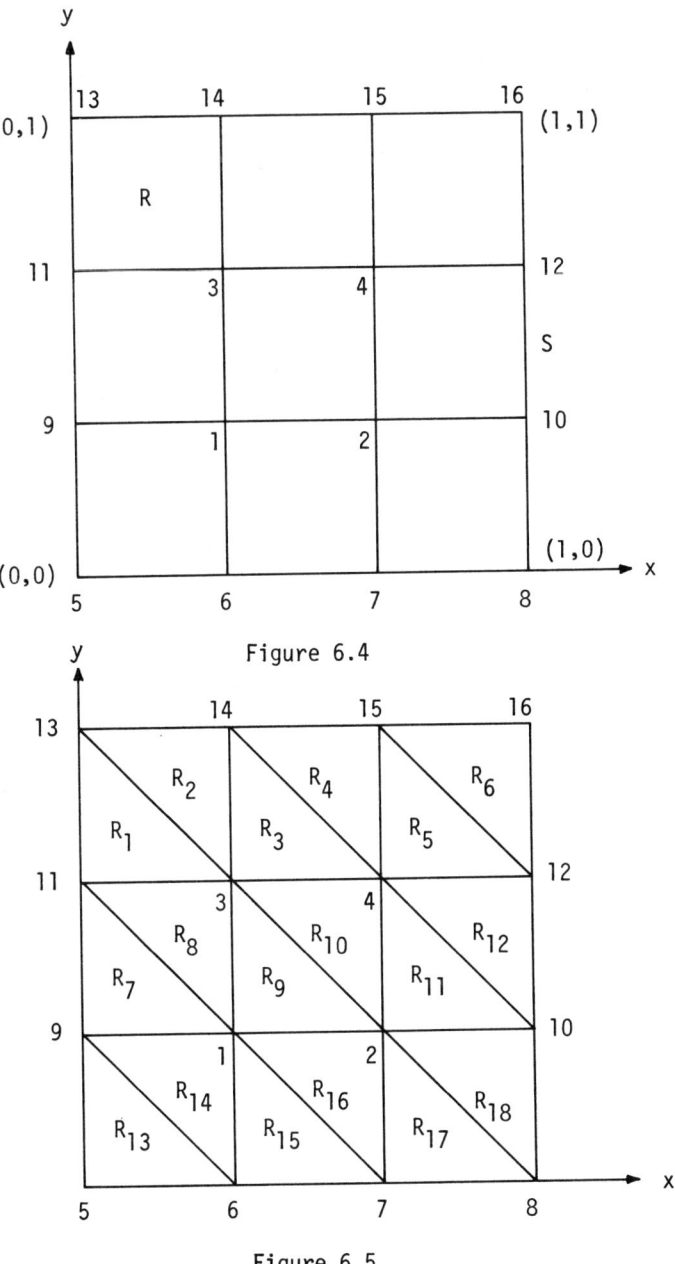

Figure 6.4

Figure 6.5

In order to approximate I_1, find the right angle vertex of S_1, which is the point numbered 11 in Figure 6.5, and at it approximate u_x and u_y by using function values only at other points of S_1. Thus,

$$u_x|_{11} \sim (u_3 - u_{11})/(1/3), \quad u_y|_{11} \sim (u_{13} - u_{11})/(1/3) ,$$

which, from (6.26) implies that

$$u_x|_{11} \sim \frac{u_3 - (-2)}{1/3} = 3u_3 + 6, \quad u_y|_{11} \sim \frac{-3 - (-2)}{1/3} = -3 .$$

Thus, I_1 can be approximated by

$$I_1^* = \frac{1}{18} \cdot \sqrt{1 + (3u_3+6)^2 + (-3)^2} .$$

Consider next

$$I_2 = \iint_{R_2+S_2} \sqrt{1 + u_x^2 + u_y^2} \, dA .$$

In order to approximate I_2, fix the right angle vertex of S_2, which is the point numbered 14 in Figure 6.5 and at it approximate u_x and u_y by using function values only at other points of S_2. Thus,

$$u_x|_{14} \sim \frac{u_{14} - u_{13}}{1/3} = \frac{-\frac{8}{3} - (-3)}{1/3} = 1$$

$$u_y|_{14} \sim \frac{u_{14} - u_3}{1/3} = \frac{-\frac{8}{3} - u_3}{1/3} = -8 - 3u_3 ,$$

and, as an approximation to I_2, take

THE PLATEAU PROBLEM

$$I_2^* = \frac{1}{18} \cdot \sqrt{1 + (1)^2 + (-8 - 3u_3)^2}.$$

Proceed in the indicated fashion until each integral

$$I_i = \iint_{R_i + S_i} \sqrt{1 + u_x^2 + u_y^2} \, dA, \quad i = 1, 2, \ldots, 18$$

is approximated by an I_i^*. In each I_i^*, u_x and u_y are approximated at the right angle vertex of S_i by means of function values only at points of S_i.

One next approximates J by J_{18}, where

$$J_{18} = \sum_{i=1}^{18} I_i^*.$$

Note immediately, that J_{18} is a function only of u_1, u_2, u_3, u_4. As an approximation to the minimum of J, we take the minimum of J_{18} at the points of R_h, and these are found, by solving the system

$$\frac{\partial J_{18}}{\partial u_i} = 0, \quad i = 1, 2, 3, 4,$$

to be

$$u_1 = -2/3, \quad u_2 = -1/3, \quad u_3 = -5/3, \quad u_4 = -4/3.$$

Interestingly enough, these values coincide with the exact values of the solution $u = x - 3y$ of the given problem at the points of R_h. But, though the given Plateau problem was somewhat trivial, the example does serve to illustrate quite simply the numerical method. For more extensive examples, see Greenspan (8).

EXERCISES

Exercises

1. Evaluate each of the functionals

   (a) $\int_0^1 [1 + (y')^2] dx$

   (b) $\int_0^1 [1 + (y')^2]^{1/2} dx$

   (c) $\int_0^1 [xy + (y')^2] dx$

   (d) $\int_0^1 [xy^4 - (y')^2 - 5xy^3 y'] dx$

   for each of the functions

   $y = x, \quad 0 \le x \le 1$

   $y = x^2, \quad 0 \le x \le 1$

   $y = x^3, \quad 0 \le x \le 1$.

2. Find the Euler equation for each functional in Exercise 1.

3. For the variational problem defined by

   $$J = \int_0^1 [1 + (y')^2] dx, \quad y(0) = 0, \; y(1) = 1,$$

   find a numerical solution with $h = 1/4$.

4. For the variational problem defined by

   $$J = \int_0^1 [1 + (y')^2]^{1/2} dx, \quad y(0) = 0, \; y(1) = 1,$$

   find a numerical solution with $h = 1/4$.

5. For the variational problem defined by

$$J = \int_0^1 [xy^2 - (y')^2 - 5xyy']dx, \quad y(0) = 0, \quad y(1) = 1,$$

find a numerical solution with $h = 1/4$.

6. In XYZ space, let S be the saddle surface whose equation is

$$x^2 - y^2 = z.$$

Find, numerically, the shortest path on S between $(-1,1,0)$ and $(4,4,0)$.

7. In XYZ space, let S be the ellipsoid whose equation is

$$x^2 + y^2 + 4z^2 = 16.$$

Find, numerically, the shortest path on the ellipsoid between $(0,0,2)$ and $(2,2,\sqrt{2})$.

8. Numerically, with $h = 1/4$, minimize the functional

$$J = \int_0^1 [y^2 + (y')^2]dx$$

subject to the boundary conditions $y(0) = 0$, $y'(1) = 1$.

9. Numerically, minimize the functional

$$J = \int_0^1 [xy + x^2y^2 + (y')^2]dx$$

subject to the boundary conditions $y(0) = 1$, $y'(1) = -1$.

EXERCISES

10. Find the Euler equation for each of the following.

(a) $J = \iint\limits_{R+S} [u_x^2 + u_y^2 + e^u] \, dA$

(b) $J = \iint\limits_{R+S} [u_x^2 + u_y^2 + u^2] \, dA$

(c) $J = \iint\limits_{R+S} [u_x^2 + u_y^2 + 2 \int_0^u H(r) \, dr] \, dA$

(d) $J = \iint\limits_{R+S} [1 + u_x^2 + u_y^2]^{1/2} \, dA$.

11. Let S be the triangle with vertices $(0,0)$, $(4,0)$, $(0,4)$. Find a numerical solution of the associated Plateau problem if $f(x,y) = x - 3y$.

12. Let S be the triangle with vertices $(0,0)$, $(4,0)$, $(0,3)$. Find a numerical solution of the associated Plateau problem if $f(x,y) = x^2 + y^2$.

13. Let S be the circle whose equation is $x^2 + y^2 = 4$. Find a numerical solution of the associated Plateau problem if $f(x,y) = x - 2y$.

CHAPTER VII

APPROXIMATE SOLUTION OF FLUID PROBLEMS

7.1 Introduction

Problems related to jet propulsion, weather prediction, molecular interaction, plasmas and flow through pipes and porous media are typical of the vast panorama of fluid problems which are of interest in science and technology. In this chapter, for convenience, we will consider only liquids and gases and prototype problems related to each. Intuitively, a liquid will be thought of as a fluid which is characterized by incompressibility and viscosity, while a gas will be thought of as a fluid which is characterized by compressibility and the absence of viscosity. Liquids will be studied first.

7.2 A Prototype Liquid Problem

A basic two dimensional, steady state, viscous, incompressible flow problem, called the cavity flow problem, can be formulated as follows. Let the points $(0,0)$, $(1,0)$, $(1,1)$, and $(0,1)$ be denoted by A, B, C, and D, respectively (see Figure 7.1). Let S be the square whose vertices are A, B, C, D and denote its interior by R. On R the equations of motion to be satisfied are the two dimensional, steady state, Navier-Stokes equations, that is,

PROTOTYPE LIQUID PROBLEM

(7.1) $$\Delta \psi = -\omega$$

(7.2) $$\Delta \omega + \mathcal{R}\left(\frac{\partial \psi}{\partial x}\frac{\partial \omega}{\partial y} - \frac{\partial \psi}{\partial y}\frac{\partial \omega}{\partial x}\right) = 0, \quad \mathcal{R} \geq 0,$$

where ψ is the stream function, ω is the vorticity, and $\mathcal{R}$ is a nonnegative constant called the Reynolds number. On S the boundary conditions to be satisfied are

(7.3) $$\psi = 0, \quad \frac{\partial \psi}{\partial x} = 0, \quad \text{on AD}$$

(7.4) $$\psi = 0, \quad \frac{\partial \psi}{\partial y} = 0, \quad \text{on AB}$$

(7.5) $$\psi = 0, \quad \frac{\partial \psi}{\partial x} = 0, \quad \text{on BC}$$

(7.6) $$\psi = 0, \quad \frac{\partial \psi}{\partial y} = -1, \quad \text{on CD}$$

The analytical problem is defined on R + S by (7.1)-(7.6) and is shown diagrammatically in Figure 7.1. Physically, one can think of a fluid contained between walls DA, AB, and BC, while a force is applied on the surface of the fluid in the direction from C to D.

Since the above problem is of wide interest to fluid dynamicists, and since the development of intuition is to be encouraged, let us, rather than merely list the algorithm, retrace the actual steps of its development. In this spirit, then, note that the first step for the numerical analyst is to immerse himself in the problem. This implies

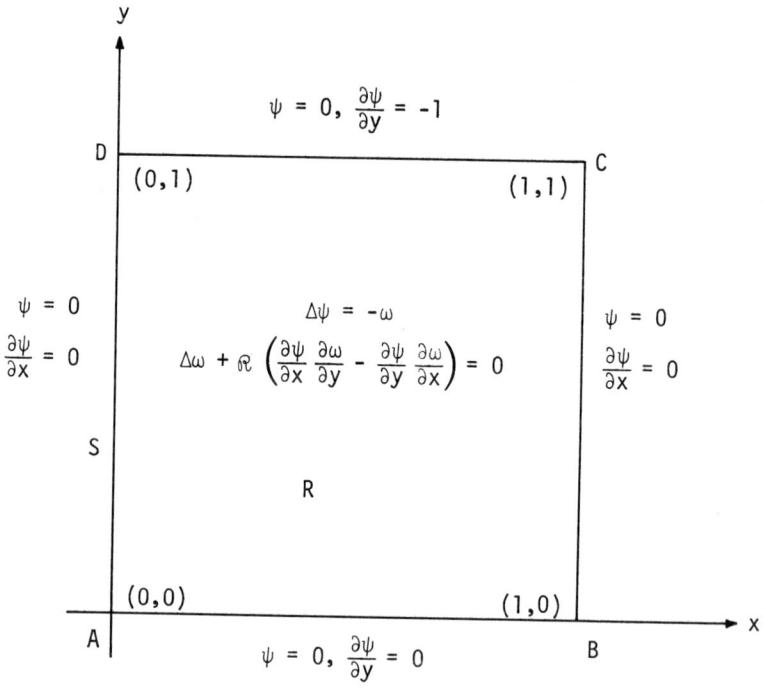

Figure 7.1

reading and discussing all aspects of the problem including existence and uniqueness theorems, experimental physical results, and available numerical methods. After two months of such activities, the following results were uncovered for the problem under consideration:

(1) Problem (7.1)-(7.6) has a unique generalized solution for small $\Re$, and at least one generalized solution for all $\Re$. (A generalized solution is one which satisfies certain integral relationships related to the differential equation, and hence need not, in general, be

differentiable everywhere). It is not known if a classical solution exists for any $R > 0$. (See Ladyzenskaya.)

(2) Many fluid dynamicists seem to feel that (7.1)-(7.6) is a reasonable approximation for the physical problem only if $R \leq 3000$, while a few aerodynamicists (see, e.g., Mills) allow $R \leq 100,000$.

(3) Laboratory experiments (Pan and Acrivos) show that for small Reynolds numbers ($R \sim 50$), the flow should look like one large vortex in the central portion of the square with two very small secondary vortices in the corners A and B. Moreover, as the Reynolds number increases, these secondary vortices disappear.

(4) Asymptotically, it has been shown (Batchelor) that as $R \to \infty$, the vorticity in a large central subregion of R converges to a constant value.

(5) Finally, several numerical methods (see Greenspan (10) for such references) had been formulated and implemented, but all were divergent for $R > 250$.

It follows, then, from the above discussion, that since we are seeking a classical solution for the problem, that is, one which is actually a solution of the differential equations on R, and since no such solution is known to exist, we must be guided in developing a numerical method by the other knowledge gathered above.

We begin with the reasoning of Kawaguti, and as a starting point in developing a method, observe that (7.1)-(7.2) is a coupled system of partial differential equations in ψ and ω. But, if ω is known, then (7.1) is a <u>linear</u> elliptic equation in ψ, while if ψ is known, then (7.2) is a <u>linear</u> elliptic equation in ω. This suggests making initial guesses $\psi^{(0)}$ and $\omega^{(0)}$ and proceeding as follows. Use $\omega^{(0)}$ in (7.1) to produce $\psi^{(1)}$. Use $\psi^{(1)}$ in (7.2) to produce $\omega^{(1)}$. Use $\omega^{(1)}$ in (7.1) to produce $\psi^{(2)}$. Use $\psi^{(2)}$ in (7.2) to produce $\omega^{(2)}$, and so on, as shown in Figure 7.2, to generate the sequences $\psi^{(0)}, \psi^{(1)}, \psi^{(2)}, \psi^{(3)}, \ldots; \omega^{(0)}, \omega^{(1)}, \omega^{(2)}, \omega^{(3)}, \ldots$. If the sequences $\psi^{(k)}$ and $\omega^{(k)}$ converge, then we might hope that they would converge to the solution ψ and ω of the given problem. Numerically, we will try to carry out such a double sequence construction on R_h and S_h, rather than on R and S, by means of the techniques developed in Chapter 3.

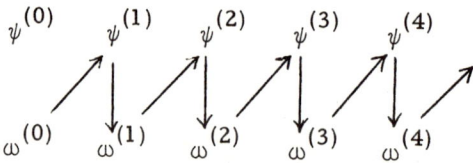

Figure 7.2

In order to proceed on R_h and S_h, we will need difference approximations for the differential equations, and, in the notation

PROTOTYPE LIQUID PROBLEM

of Figure 7.3, we begin with the following. If ω is considered to be known, then approximate (7.1) by

(7.7)
$$\frac{-4\psi_0 + \psi_1 + \psi_2 + \psi_3 + \psi_4}{h^2} = -\omega_0 ,$$

while, if ψ is considered to be known, approximate (7.2) by

(7.8)
$$\frac{-4\omega_0 + \omega_1 + \omega_2 + \omega_3 + \omega_4}{h^2} + \mathcal{R}\left(\frac{\psi_1-\psi_3}{2h}\cdot\frac{\omega_2-\omega_4}{2h} - \frac{\psi_2-\psi_4}{2h}\cdot\frac{\omega_1-\omega_3}{2h}\right) = 0,$$

or, equivalently, by

(7.9)
$$-4\omega_0 + [1 - \frac{\mathcal{R}}{4}(\psi_2 - \psi_4)]\omega_1 + [1 + \frac{\mathcal{R}}{4}(\psi_1 - \psi_3)]\omega_2$$
$$+ [1 + \frac{\mathcal{R}}{4}(\psi_2 - \psi_4)]\omega_3 + [1 - \frac{\mathcal{R}}{4}(\psi_1 - \psi_3)]\omega_4 = 0 .$$

Observe next that to generate ψ on R_h from (7.7), one must know ψ on S_h, and this is available from (7.3)-(7.6). But, to generate ω on R_h from (7.9), one must know ω on S_h, and this is <u>not</u> available from (7.3)-(7.6). Thus, when generating ψ, one need only be concerned with R_h, but when generating ω, one must be concerned with both R_h and S_h. Since (7.9) is proposed to be used for R_h, we must decide how to approximate ω on S_h, and this will be done as follows.

Assume that (7.1) is satisfied on S and consider the Laplace operator $\psi_{xx} + \psi_{yy}$. Let (x,y), $(x+h,y)$, $(x,y+h)$, $(x,y-h)$ be numbered

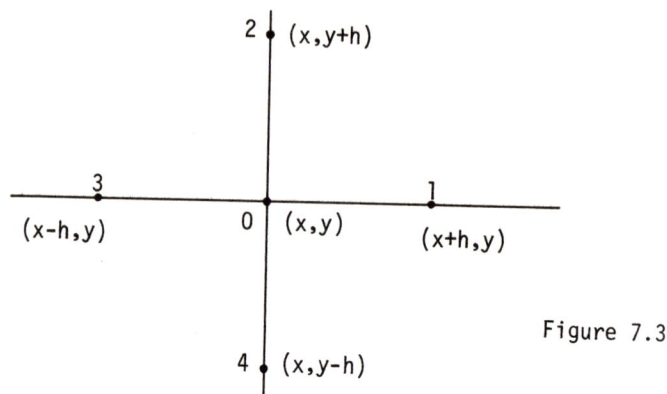

Figure 7.3

0, 1, 2, 4, respectively, as shown in Figure 7.4a. Let 0, 2 and 4 be in S, while 1 is in R. Consider the determination of parameters $\alpha_0, \alpha_1, \alpha_2, \alpha_4, \alpha_5$ such that

(7.10) $\quad (\psi_{xx} + \psi_{yy})|_0 = \alpha_0 \psi_0 + \alpha_1 \psi_1 + \alpha_2 \psi_2 + \alpha_4 \psi_4 + \alpha_5 \left(\dfrac{\partial \psi}{\partial x}\right)\Big|_0 .$

In (7.10) expansion of ψ_1, ψ_2 and ψ_4 into Taylor series about the point numbered 0 and reorganization of terms implies

$$(\psi_{xx} + \psi_{yy})|_0 = \psi_0(\alpha_0 + \alpha_1 + \alpha_2 + \alpha_4) + \psi_x(h\alpha_1 + \alpha_5) + \psi_y(h\alpha_2 - h\alpha_4)$$
$$+ \psi_{xx}\left(\dfrac{h^2}{2}\alpha_1\right) + \psi_{yy}\left(\dfrac{h^2}{2}\alpha_2 + \dfrac{h^2}{2}\alpha_4\right) + \cdots$$

In this latter equality, the setting of corresponding coefficients equal yields

$$\alpha_0 + \alpha_1 + \alpha_2 + \alpha_4 = 0$$
$$h\alpha_1 + \alpha_5 = 0, \quad h\alpha_2 - h\alpha_4 = 0$$
$$\dfrac{h^2}{2}\alpha_1 = 1, \quad \dfrac{h^2}{2}\alpha_2 + \dfrac{h^2}{2}\alpha_4 = 1,$$

PROTOTYPE LIQUID PROBLEM

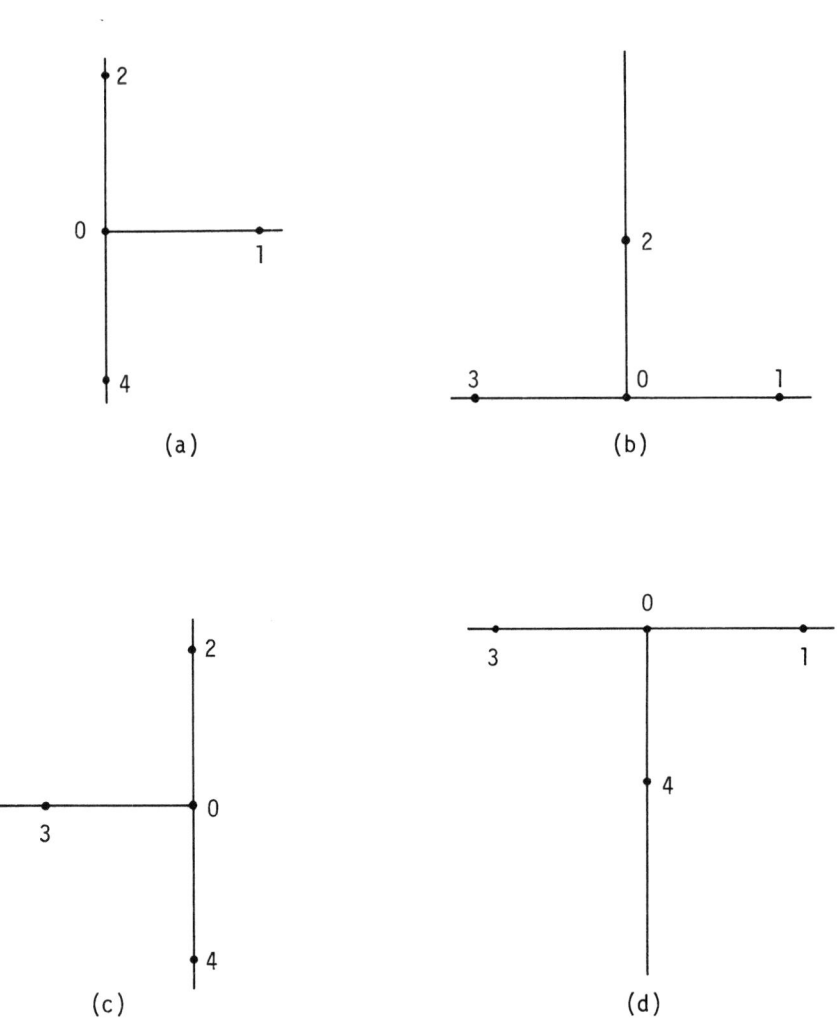

Figure 7.4

the solution of which is

$$\alpha_0 = -\frac{4}{h^2}, \quad \alpha_1 = \frac{2}{h^2}, \quad \alpha_2 = \alpha_4 = \frac{1}{h^2}, \quad \alpha_5 = -\frac{2}{h}.$$

Thus one arrives at the following approximation:

(7.11) $\quad (\psi_{xx} + \psi_{yy})|_0 = -\frac{4}{h^2}\psi_0 + \frac{2}{h^2}\psi_1 + \frac{1}{h^2}\psi_2 + \frac{1}{h^2}\psi_4 - \frac{2}{h}\left(\frac{\partial\psi}{\partial x}\right)\bigg|_0.$

Finally, from (7.1), (7.3) and (7.11), it follows that

(7.12) $\quad\quad\quad\quad\quad\quad \omega_0 = -\frac{2\psi_1}{h^2}.$

Similarly, for the point arrangement shown in Figure 7.4b,

(7.13) $\quad\quad\quad\quad\quad\quad \omega_0 = -\frac{2\psi_2}{h^2};$

for that in Figure 7.4c,

(7.14) $\quad\quad\quad\quad\quad\quad \omega_0 = -\frac{2\psi_3}{h^2};$

while, for Figure 7.4d, since $\psi_y = -1$ on DC,

(7.15) $\quad\quad\quad\quad\quad\quad \omega_0 = \frac{2}{h} - \frac{2\psi_4}{h^2}.$

An algorithm can now be formulated as follows.

Method N. S.

Step 1 For given h, construct and number the points of R_h and S_h as is usual for elliptic problems.

PROTOTYPE LIQUID PROBLEM 217

Step 2   Set $\psi_i^{(0)} = 0$ on R_h and $\omega_i^{(0)} = 0$ on $R_h + S_h$.

Step 3   Generate sequences $\psi^{(k)}$ on R_h and $\omega^{(k)}$ on $R_h + S_h$ by the alternating procedure shown in Figure 7.2 as follows. Apply method D with

(7.16) $\qquad -4\psi_0^{(k)} + \psi_1^{(k)} + \psi_2^{(k)} + \psi_3^{(k)} + \psi_4^{(k)} = -h^2 \omega_0^{(k-1)}, \quad k = 1, 2, \ldots,$

to generate $\psi^{(k)}$ on R_h. Apply then

(7.17) $\qquad \omega_0^{(k)} = -2\psi_1^{(k)} / (h^2)$

on AD;

(7.18) $\qquad \omega_0^{(k)} = -2\psi_2^{(k)} / (h^2)$

on AB;

(7.19) $\qquad \omega_0^{(k)} = -2\psi_3^{(k)} / (h^2)$

on BC;

(7.20) $\qquad \omega_0^{(k)} = \frac{2}{h} - \frac{2\psi_4^{(k)}}{h^2}$

on CD, to approximate $\omega^{(k)}$ on S_h. Then, apply Method D with

(7.21) $\quad -4\omega_0^{(k)} + [1 - \frac{R}{4}(\psi_2^{(k)} - \psi_4^{(k)})]\omega_1^{(k)} + [1 + \frac{R}{4}(\psi_1^{(k)} - \psi_3^{(k)})]\omega_2^{(k)}$

$\qquad + [1 + \frac{R}{4}(\psi_2^{(k)} - \psi_4^{(k)})]\omega_3^{(k)} + [1 - \frac{R}{4}(\psi_1^{(k)} - \psi_3^{(k)})]\omega_4^{(k)} = 0$

to generate $\omega^{(k)}$ on R_h.

__Step 4__ For given positive ε_1 and ε_2, terminate the iteration of Step 3 when

(7.22) $$|\psi^{(k)} - \psi^{(k+1)}| < \varepsilon_1 \, , \quad \text{uniformly on } R_h$$

(7.23) $$|\omega^{(k)} - \omega^{(k+1)}| < \varepsilon_2 \, , \quad \text{uniformly on } R_h + S_h \, .$$

__Step 5__ Call $\psi^{(k+1)}$ and $\omega^{(k+1)}$, so generated, the numerical solution of the given problem.

Method N.S. works relatively well but diverges for $\mathcal{R} > 250$ and $h = \frac{1}{40}$. However, rather than give the numerical results so derived, let us delay and give the results after the method has been modified to eliminate the divergent behavior.

In order to probe the reasons for divergence when $\mathcal{R} > 250$, the computer output had to be read item by item and very often graphed by hand so as not to miss possible troubles in the program and/or the method. After a week of such study of the massive computer output, it was found that divergence resulted because the generalized Newton's method was diverging in its attempt to generate $\omega^{(k)}$. This led to an actual listing of the equation (7.21) for each value of k and it was discovered that diagonal dominance had been lost. Indeed, the terms $\frac{\mathcal{R}}{4}(\psi_2^{(k)} - \psi_4^{(k)})$ and $\frac{\mathcal{R}}{4}(\psi_1^{(k)} - \psi_3^{(k)})$ were becoming so large that the matrix of the resulting system was losing its diagonal dominance. The

PROTOTYPE LIQUID PROBLEM

natural remedy then, to maintain the diagonal dominance, was to introduce the forward-backward technique described in Section 2.8, and this was done as follows.

Modification 1 of Method N.S.

Set

$$\alpha = \psi_1^{(k)} - \psi_3^{(k)}$$

$$\beta = \psi_2^{(k)} - \psi_4^{(k)}$$

and approximate (7.2) by

$$\frac{-4\omega_0^{(k)} + \omega_1^{(k)} + \omega_2^{(k)} + \omega_3^{(k)} + \omega_4^{(k)}}{h^2} + \frac{\Re \alpha A}{2h} - \frac{\Re \beta B}{2h} = 0,$$

where

$$A = \frac{\omega_2^{(k)} - \omega_0^{(k)}}{h}, \text{ if } \alpha \geq 0; \quad A = \frac{\omega_0^{(k)} - \omega_4^{(k)}}{h}, \text{ if } \alpha < 0$$

$$B = \frac{\omega_0^{(k)} - \omega_3^{(k)}}{h}, \text{ if } \beta \geq 0; \quad B = \frac{\omega_1^{(k)} - \omega_0^{(k)}}{h}, \text{ if } \beta < 0,$$

or, equivalently, replace (7.21) by

(7.24) $\quad (-4 - \frac{\alpha \Re}{2} - \frac{\beta \Re}{2}) \omega_0^{(k)} + \omega_1^{(k)} + (1 + \frac{\alpha \Re}{2}) \omega_2^{(k)} + (1 + \frac{\beta \Re}{2}) \omega_3^{(k)}$

$$+ \omega_4^{(k)} = 0, \text{ if } \alpha \geq 0, \beta \geq 0,$$

(7.25) $\quad (-4 - \frac{\alpha \Re}{2} + \frac{\beta \Re}{2}) \omega_0^{(k)} + (1 - \frac{\beta \Re}{2}) \omega_1^{(k)} + (1 + \frac{\alpha \Re}{2}) \omega_2^{(k)} + \omega_3^{(k)}$

$$+ \omega_4^{(k)} = 0, \text{ if } \alpha \geq 0, \beta < 0,$$

$$(7.26) \quad (-4 + \frac{\alpha R}{2} - \frac{\beta R}{2})\omega_0^{(k)} + \omega_1^{(k)} + \omega_2^{(k)} + (1 + \frac{\beta R}{2})\omega_3^{(k)}$$

$$+ (1 - \frac{\alpha R}{2})\omega_4^{(k)} = 0, \quad \text{if} \quad \alpha < 0, \quad \beta \geq 0,$$

$$(7.27) \quad (-4 + \frac{\alpha R}{2} + \frac{\beta R}{2})\omega_0^{(k)} + (1 - \frac{\beta R}{2})\omega_1^{(k)} + \omega_2^{(k)} + \omega_3^{(k)}$$

$$+ (1 - \frac{\alpha R}{2})\omega_4^{(k)} = 0, \quad \text{if} \quad \alpha < 0, \quad \beta < 0.$$

The application of the new algorithm resulted in convergence for $0 \leq R \leq 10^5$, $h = \frac{1}{10}$, but diverged for $h < \frac{1}{10}$. So, the output had to be studied very carefully again, and this time it was discovered that the numerical results were not good approximations for the derivative conditions on S_h, and that $\omega^{(k)}$ was diverging on S_h. To remedy the situation, since both deficiencies were on the boundary, a set of "inner boundary" points like those shown to be crossed in Figure 7.5, was introduced and the method was modified as follows to force good numerical approximations to the derivative conditions on the boundary.

At a point (x,y) of AD and numbered 0, as in Figure 7.6a set

$$(7.28) \quad \left.\frac{\partial \psi}{\partial x}\right|_0 = \frac{1}{2h}(-3\psi_0 + 4\psi_1 - \psi_2).$$

By (7.3), then, (7.28) reduces to

$$(7.29) \quad \psi_1 = \frac{\psi_2}{4}.$$

PROTOTYPE LIQUID PROBLEM   221

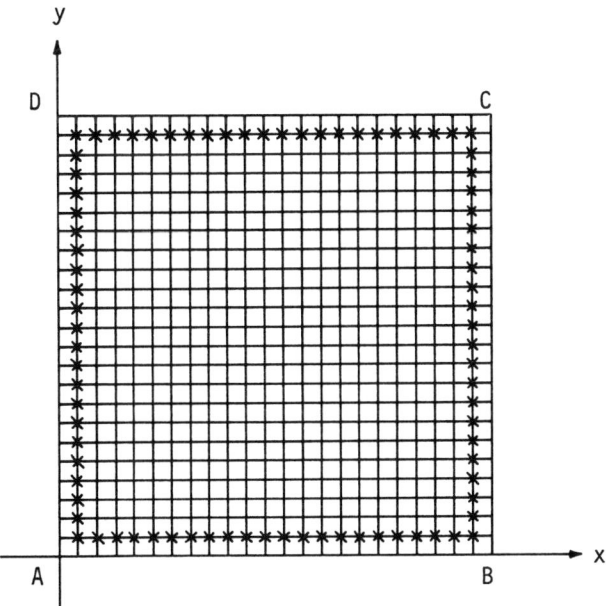

Figure 7.5

If one uses a similar numbering of points on the other parts of S_h, as shown in Figures 7.6(b), (c), then in each case (7.29) follows, while for Figure 7.5(d), because $\psi_y = -1$ on DC, one has

$$\psi_1 = \frac{h}{2} + \frac{\psi_2}{4}.$$

Thus, the following modification was made.

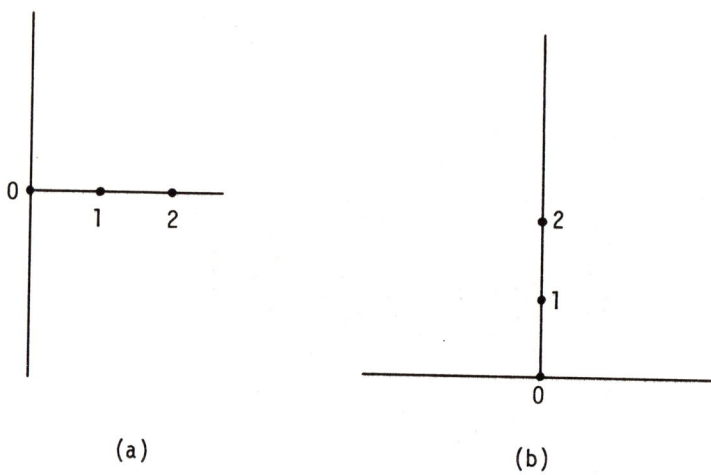
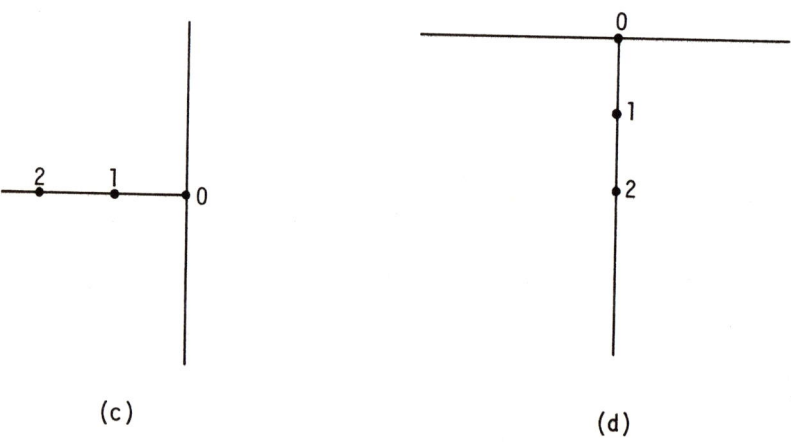

Figure 7.6

Modification 2 of Method N.S.

At each point of R_h of the form (h, ih), $i = 2, 3, \ldots, n-2$, set

(7.30) $$\psi^{(k)}(h, ih) = \frac{\psi^{(k)}(2h, ih)}{4} .$$

At each point of R_h of the form (ih, h), $i = 1, 2, \ldots, n-1$, set

(7.31) $$\psi^{(k)}(ih, h) = \frac{\psi^{(k)}(ih, 2h)}{4} .$$

At each point of R_h of the form $(1-h, ih)$, $i = 2, 3, \ldots, n-2$, set

(7.32) $$\psi^{(k)}(1-h, ih) = \frac{\psi^{(k)}(1-2h, ih)}{4} .$$

At each point of R_h of the form $(ih, 1-h)$, $i = 1, 2, \ldots, n-1$, set

(7.33) $$\psi^{(k)}(ih, 1-h) = \frac{h}{2} + \frac{\psi^{(k)}(ih, 1-2h)}{4} .$$

Then, apply (7.16) only on the remaining points of R_h and solve the resulting system to generate $\psi^{(k)}$ on R_h.

Note, incidentally, that the application of (7.30)-(7.33) on the inner boundary preserves diagonal dominance.

Computer runs with the new algorithm resulted in convergence for $0 \leq \Re \leq 10^5$ and $h = \frac{1}{15}$, but divergence for $h < \frac{1}{15}$. Detailed study of the output revealed that sequence $\omega^{(k)}$ was still diverging on S_h. But analysis of the divergence showed that, at certain points of S_h, $\omega^{(k)}$ seemed to be converging and then suddenly it would overstep its limit and become unbounded rapidly. Thus, it was deemed reasonable to slow down the convergence rate of $\omega^{(k)}$, and this was done as follows.

Modification 3 of Method N.S.

On $R_h + S_h$, denote the resulting approximation by Step 3 in Method N.S. of ω by $\bar{\omega}^{(k)}$. Then, at each point of $R_h + S_h$ define $\omega^{(k)}$ by the smoothing formula

(7.34) $$\omega^{(k)} = \mu \omega^{(k-1)} + (1-\mu)\bar{\omega}^{(k)}, \quad 0 \leq \mu \leq 1.$$

Computer examples this time converged for $0 \leq R \leq 10^5$ and $h = \frac{1}{20}$, but diverged for $h < \frac{1}{20}$. The divergence again was the result of $\omega^{(k)}$ becoming unbounded at points of S_h.

Finally, for no reason other than all other possible modifications seemed to have been exhausted, it was decided to explore a smoothing of ψ in the following fashion.

Modification 4 of Method N.S.

On R_h, denote the resulting approximation by Step 3 in Method N.S. of ψ by $\bar{\psi}^{(k)}$. Then, at each point of R_h, define $\psi^{(k)}$ by

(7.35) $$\psi^{(k)} = \rho \psi^{(k-1)} + (1-\rho)\bar{\psi}^{(k)}, \quad 0 \leq \rho \leq 1.$$

Computer examples then converged for $0 \leq R \leq 10^5$ and, in order, for $h = \frac{1}{25}, \frac{1}{30}, \frac{1}{40}, \frac{1}{50}, \frac{1}{60}, \frac{1}{70}, \frac{1}{80}$, and $\frac{1}{99}$, at which

PROTOTYPE LIQUID PROBLEM 225

point storage problems became significant and the computations ceased. Let us then show next some of the numerical results, which are given as continuous curves by interpolation on the grid lines.

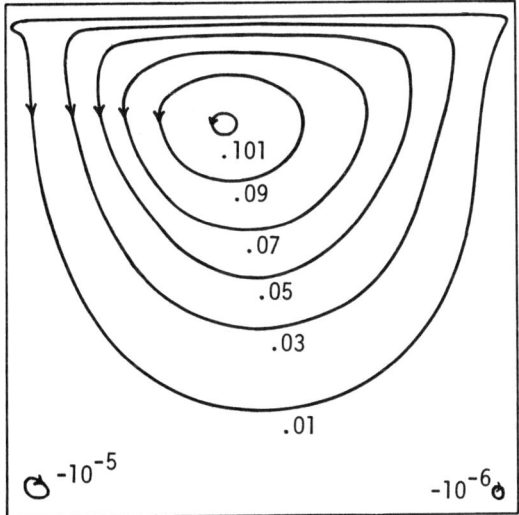

Figure 7.7(a)

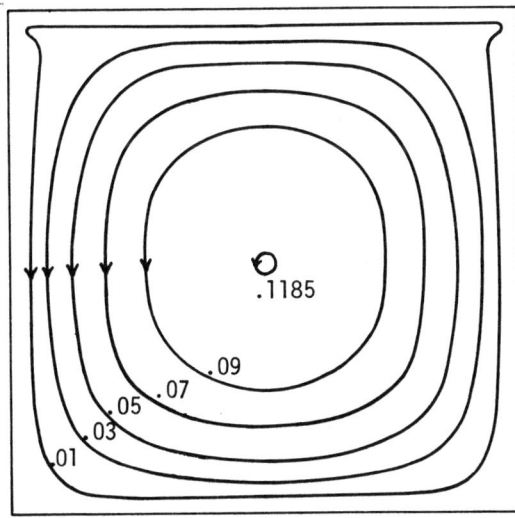

Figure 7.7(b)

Level ψ curves, called streamlines, are shown in Figure 7.7a, where a primary and two secondary vortices are shown for $\Re = 50$, $h = \frac{1}{40}$, $\varepsilon_1 = \varepsilon_2 = 10^{-3}$, $\rho = 0.03$, $\mu = 0.90$. The running time on the UNIVAC 1108 was 28 minutes. Streamlines are shown in Figure 7.7b for $\Re = 10^5$, $h = \frac{1}{40}$, $\varepsilon_1 = \varepsilon_2 = 10^{-4}$, $\rho = 0.03$, $\mu = 0.95$. Convergence was achieved in 27 minutes, but the starting values $\psi^{(0)}$ and $\omega^{(0)}$ were taken from a previous computation with $\Re = 10^4$ (see Greenspan (8)). The disappearance of the secondary vortices in Figure 7.7b is in agreement with the experimental results of Pan and Acrivos. In Figure 7.8 is shown for the case $\Re = 10^5$ the level curve $\omega = 1.630$, with its double spiral, space filling character. Numerical verification of Batschelor's result that vorticity in a large central subregion of R converges to a constant as $\Re \to \infty$ is exhibited in Figure 7.8 by a darkening of those points at which the vorticity is between 1.6 and 1.7.

A final note of interest is in order. Though the residuals were slightly higher, the numerical results using (7.24)-(7.27) also proved to be a solution of (7.21) for all values of $\Re$ in the range studied.

BIHARMONIC PROBLEMS

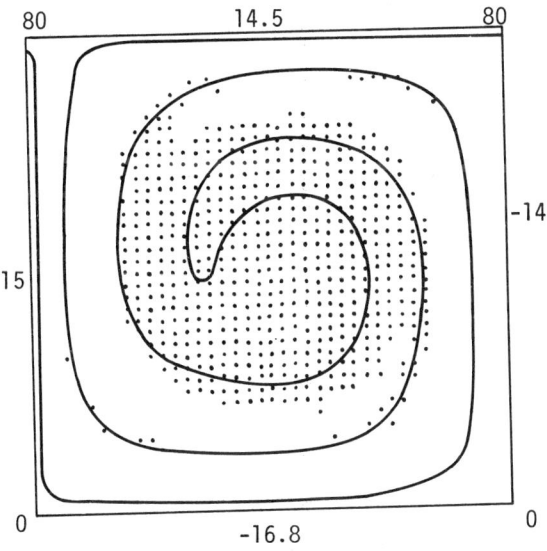

Figure 7.8

7.3 Biharmonic Problems

At this point it is convenient to notice that a variety of problems in both fluid dynamics and elasticity can be formulated as biharmonic problems, which, in terms of the square of Figure 7.1, can be stated as follows. Find a solution on R of the biharmonic equation

(7.36) $$\frac{\partial^4 \psi}{\partial x^4} + 2 \frac{\partial^4 \psi}{\partial x^2 \partial y^2} + \frac{\partial^4 \psi}{\partial y^4} = 0,$$

given ψ on S, given ψ_x on AD and BC, and given ψ_y on CD and AB. If one notes that setting

(7.37) $$\frac{\partial^2 \psi}{\partial x^2} + \frac{\partial^2 \psi}{\partial y^2} \equiv -\omega$$

implies

(7.38) $$\frac{\partial^2 \omega}{\partial x^2} + \frac{\partial^2 \omega}{\partial y^2} \equiv -\left(\frac{\partial^4 \psi}{\partial x^4} + 2\frac{\partial^4 \psi}{\partial x^2 \partial y^2} + \frac{\partial^4 \psi}{\partial y^4}\right),$$

then it follows readily that the biharmonic equation is equivalent to the system

(7.39) $$\Delta \psi = -\omega$$

(7.40) $$\Delta \omega = 0.$$

But this system is a special case of (7.1)-(7.2) with $\mathcal{R} = 0$, and can then be solved easily by the method of Section 7.2.

7.4 A Prototype Time Dependent Fluid Problem

Thus far attention has been restricted to <u>steady state</u>, liquid-type problems. Attention will be directed next to <u>nonsteady</u>, or <u>time dependent</u>, liquid type problems and will be restricted, for illustrative purposes, only to the following prototype cavity problem. Find two functions $\psi(x,y,t)$ and $\omega(x,y,t)$, $0 \leq x \leq 1$, $0 \leq y \leq 1$, $0 \leq t$ such that for a given positive number $\mathcal{R}$ the following are valid:

(7.41) $$\Delta \psi \equiv \frac{\partial^2 \psi}{\partial x^2} + \frac{\partial^2 \psi}{\partial y^2} = -\omega; \quad 0 < x < 1, \ 0 < y < 1, \ 0 < t$$

(7.42) $\quad \dfrac{\partial \omega}{\partial t} = \dfrac{1}{\mathcal{R}} \Delta \omega + \dfrac{\partial \psi}{\partial x}\dfrac{\partial \omega}{\partial y} - \dfrac{\partial \psi}{\partial y}\dfrac{\partial \omega}{\partial x};\quad 0 < x < 1,\ 0 < y < 1,\ 0 < t$

(7.43) $\quad \psi(x,y,0) = \omega(x,y,0) = 0;\quad 0 \le x \le 1,\ 0 \le y \le 1$

(7.44) $\quad \psi(x,0,t) = \dfrac{\partial \psi(x,0,t)}{\partial y} = 0;\quad 0 \le x \le 1,\ 0 \le t$

(7.45) $\quad \psi(0,y,t) = \dfrac{\partial \psi(0,y,t)}{\partial x} = \psi(1,y,t) = \dfrac{\partial \psi(1,y,t)}{\partial x} = 0;\quad 0 \le y \le 1,\ 0 \le t$

and

(7.46) $\quad \psi(x,1,t) = 0,\ \dfrac{\partial \psi}{\partial y}(x,1,t) = -1;\quad 0 \le x \le 1,\ 0 \le t.$

7.5 A Boundary Value Technique

Because of its simplicity, economy, and similarity to the methods discussed thus far, a boundary value technique for problem (7.41)-(7.46) will be developed first. Application of this method, unlike those which follow, requires the assumption that (7.41)-(7.46) has a steady state solution which is independent of the initial conditions. The method proceeds as follows.

With grid size h, first solve numerically the steady state problem (7.1)-(7.6). This numerical solution is denoted by $\psi_h(x,y,\infty)$, $\omega_h(x,y,\infty)$. Next, for fixed $t = T$, subdivide $0 \le t \le T$ into n equal parts, each of length Δt, by means of the points $t_k = k\Delta t$, $k = 0,1,\ldots,n$, and, as shown in Figure 7.9, let P be the rectangular parallelepiped defined by $P = \{(x,y,t): 0 \le x \le 1,\ 0 \le y \le 1,\ 0 \le t \le T\}$.

230   FLUID PROBLEMS

Define R to be the interior and S to be the boundary of P.

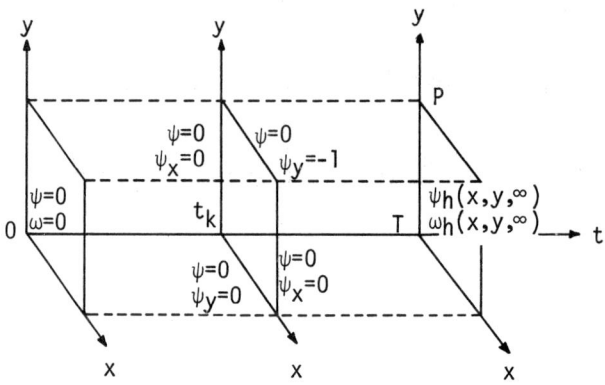

Figure 7.9

Using space grid size h and time grid size Δt, then construct in the usual way the three dimensional sets of interior grid points, denoted this time by $R_{h,\Delta t}$, and boundary grid points, denoted this time by $S_{h,\Delta t}$. At the boundary grid points set

(7.47) $\quad \psi(x,y,0) = 0$

(7.48) $\quad \psi(x,y,T) = \psi(x,y,\infty)$

(7.49) $\quad \psi(0,y,t) = \psi(1,y,t) = \psi(x,0,t) = \psi(x,1,t) = 0$.

At each point (x,y,t) of $R_{h,\Delta t}$ define

(7.50) $\quad \psi^{(0)}(x,y,t) = 0, \ \omega^{(0)}(x,y,t) = 0$.

BOUNDARY VALUE TECHNIQUE

We now show how to construct from (7.50) on $R_{h,\Delta t}$ a sequence of discrete functions

(7.51) $$\psi^{(1)}, \psi^{(2)}, \psi^{(3)}, \ldots, \psi^{(k)}$$

and on $R_{h,\Delta t} + S_{h,\Delta t}$ a sequence of discrete functions

(7.52) $$\omega^{(1)}, \omega^{(2)}, \omega^{(3)}, \ldots, \omega^{(k)}$$

which will both converge. For this purpose, at each point of $R_{h,\Delta t}$ write down the difference equation

(7.53) $$-4\psi(x,y,t_k) + \psi(x+h,y,t_k) + \psi(x,y+h,t_k) + \psi(x-h,y,t_k)$$
$$+ \psi(x,y-h,t_k) = -h^2 \omega^{(0)}(x,y,t_k), \quad k = 1,2,\ldots,n-1.$$

Inserting the known values (7.47)-(7.49) wherever possible into (7.53), solve the resulting linear algebraic system by the generalized Newton's method. If the resulting solution is denoted by $\bar{\psi}^{(1)}$, then on $R_{h,\Delta t}$ the element $\psi^{(1)}$ of (7.51) is defined by the smoothing formula

(7.54) $$\psi^{(1)} = \rho \psi^{(0)} + (1-\rho)\bar{\psi}^{(1)}, \quad 0 \le \rho \le 1.$$

Next, at each point of $S_{h,\Delta t}$ set

(7.55) $$\bar{\omega}^{(1)} = \frac{2\xi}{h} - \frac{2\psi^{(1)}}{h^2},$$

where $\psi^{(1)}$ is evaluated at the point of $R_{h,\Delta t}$ which is nearest to the given point and where

(7.56) $$\xi = \begin{cases} 1, & \text{if } y = 1 \\ 0, & \text{if } y \neq 1. \end{cases}$$

Note that (7.55) is a concise way of expressing (7.12)-(7.15). At each point of $R_{h,\Delta t}$, write down the difference equation

(7.57) $$\frac{1}{2\Delta t}[-3\omega(x,y,t-\Delta t) + 4\omega(x,y,t) - \omega(x,y,t+\Delta t)]$$

$$= \frac{1}{h^2 \mathcal{R}}[-4\omega(x,y,t) + \omega(x+h,y,t) + \omega(x,y+h,t)$$

$$+ \omega(x-h,y,t) + \omega(x,y-h,t)] + \alpha \cdot \beta - \gamma \cdot \delta,$$

where

(7.58) $$\alpha = \frac{\psi^{(1)}(x+h,y,t) - \psi^{(1)}(x-h,y,t)}{2h}$$

(7.59) $$\gamma = \frac{\psi^{(1)}(x,y+h,t) - \psi^{(1)}(x,y-h,t)}{2h}$$

(7.60) $$\beta = \frac{\omega(x,y+h,t) - \omega(x,y,t)}{h}, \quad \text{if } \alpha \geq 0$$

(7.61) $$\beta = \frac{\omega(x,y,t) - \omega(x,y-h,t)}{h}, \quad \text{if } \alpha < 0$$

(7.62) $$\delta = \frac{\omega(x,y,t) - \omega(x-h,y,t)}{h}, \quad \text{if } \gamma \geq 0$$

(7.63) $$\delta = \frac{\omega(x+h,y,t) - \omega(x,y,t)}{h}, \quad \text{if } \gamma < 0.$$

BOUNDARY VALUE TECHNIQUE

Insert the values (7.55) wherever possible and solve the system generated by (7.57). The solution is denoted by $\bar{\omega}^{(1)}$. Then, on $R_{h,\Delta t} + S_{h,\Delta t}$ define

(7.64) $\qquad \omega^{(1)} = \mu \omega^{(0)} + (1-\mu)\bar{\omega}^{(1)}, \quad 0 \leq \mu \leq 1$,

which completes the construction of the element $\omega^{(1)}$ in sequence (7.52).

One proceeds next to construct $\psi^{(2)}$ from $\psi^{(1)}$ and $\omega^{(1)}$ in the same spirit as $\psi^{(1)}$ was constructed from $\psi^{(0)}$ and $\omega^{(0)}$, and to generate $\omega^{(2)}$ from $\psi^{(2)}$ just as $\omega^{(1)}$ was generated from $\psi^{(1)}$. In the indicated fashion, the iteration continues until, for some preassigned tolerances ε_1 and ε_2, one finds that, uniformly,

(7.65) $\qquad |\psi^{(k)} - \psi^{(k+1)}| < \varepsilon_1$

(7.66) $\qquad |\omega^{(k)} - \omega^{(k+1)}| < \varepsilon_2$.

The discrete functions $\psi^{(k+1)}$ and $\omega^{(k+1)}$ are taken to be numerical approximations of $\psi(x,y,t)$ and $\omega(x,y,t)$, respectively.

Various examples for $0 < \Re \leq 500$ were run on the UNIVAC 1108. Since all behaved similarly, we shall discuss in detail only the case $\Re = 500$, with which other authors have found exceptional difficulties (see, e.g., Pearson).

The steady state problem (7.1)-(7.6) was solved by the method of Section 7.2 for $h = \frac{1}{20}$. These results are shown graphically in Figures 7.10 and 7.11. Then, with $T = 5$, $\Delta t = \frac{1}{2}$, $\rho = 0.3$, $\mu = 0.7$ and $\varepsilon_1 = \varepsilon_2 = 10^{-3}$, the method of this section converged in 54 iterations, which took 8 minutes of running time. The stream curves $\psi = 0.09, 0.07, 0.05, 0.03, 0.01$, and equivorticity curves $\omega = 4, 1.6, 0$ are given for $t = 1,2,3,4$ in Figures (7.12)-(7.19).

It should be observed from this example that one gets excellent results for <u>large</u> time steps, which makes the method relatively attractive.

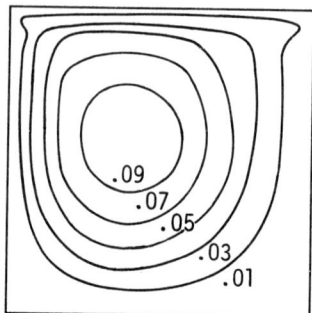
Steady state stream curves.

Figure 7.10

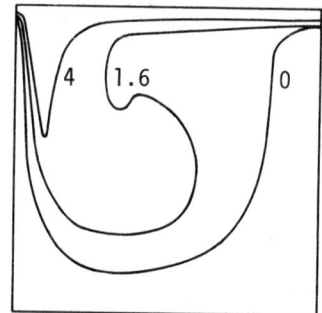
Steady state equivorticity curves.

Figure 7.11

BOUNDARY VALUE TECHNIQUE 235

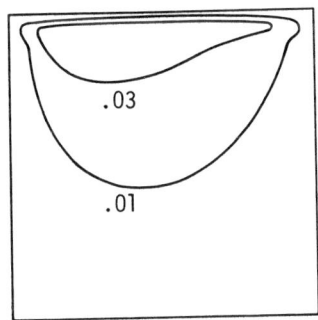

Streamlines at t=1.

Figure 7.12

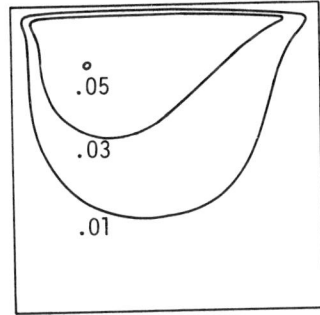

Streamlines at t=2.

Figure 7.13

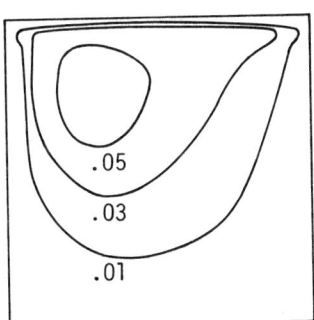

Streamlines at t=3.

Figure 7.14

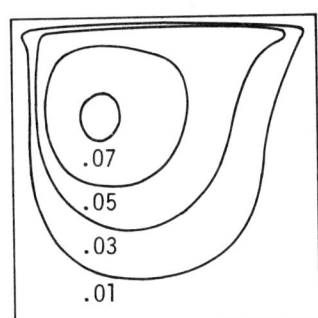

Streamlines at t=4.

Figure 7.15

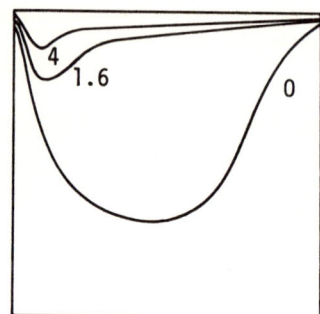

Equivorticity curves at t=1.

Figure 7.16

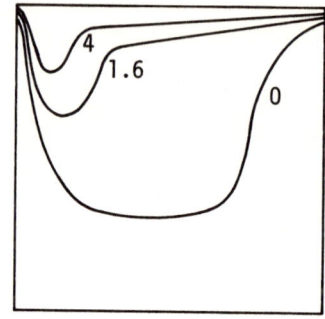

Equivorticity curves at t=2.

Figure 7.17

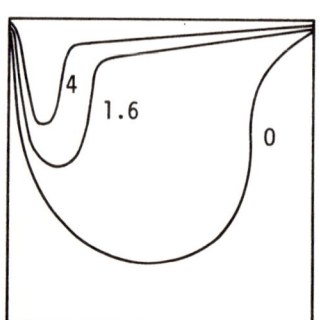

Equivorticity curves at t=3.

Figure 7.18.

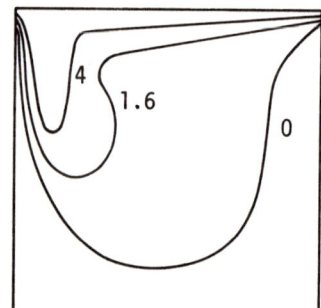

Equivorticity curves at t=4.

Figure 7.19

7.6 The Method of Fromm

An explicit, step-by-step method which has proved to be of value for a large class of time dependent problems, whether or not a steady state solution exists, is the method of Fromm. The basic idea is to preserve certain energy relationships (see Fromm) by differencing (7.64) in the equivalent form

(7.67) $$\frac{\partial \omega}{\partial t} = \frac{1}{R} \Delta \omega + [\frac{\partial}{\partial y}(\psi_x \omega) - \frac{\partial}{\partial x}(\psi_y \omega)].$$

As applied to Problem (7.41)-(7.46), the method can be given as follows. Let Δt and $\Delta x = \Delta y = h$ be grid sizes, and, for simplicity, use the notation

(7.68) $$u_{i,j} = [\psi_{i,j+1} - \psi_{i,j-1}]/[2h]$$

(7.69) $$v_{i,j} = [\psi_{i+1,j} - \psi_{i-1,j}]/[2h]$$

(7.70) $$\psi_{i,j}^{(k)} = \psi(ih, jh, k\Delta t)$$

(7.71) $$\omega_{i,j}^{(k)} = \omega(ih, jh, k\Delta t).$$

Then,

Step 1 Generate $\omega_{i,j}^{(1)}$ at interior grid points by any standard method. One can do this by applying any of the methods for parabolic problems to either (7.42) or (7.67). Then, generate $\psi_{i,j}^{(1)}$ by applying Method D with the difference equation

(7.72) $\quad -4\psi_{i,j}^{(1)} + \psi_{i+1,j}^{(1)} + \psi_{i,j+1}^{(1)} + \psi_{i-1,j}^{(1)} + \psi_{i,j-1}^{(1)} = -h^2 \omega_{i,j}^{(1)}$.

Finally, determine $\omega_{i,j}^{(1)}$ at boundary grid points by means of (7.55).

<u>Step 2</u> To proceed, in general, from t_k to t_{k+1}, $k = 1, 2, \ldots$, generate $\omega_{i,j}^{(k+1)}$ explicitly at interior grid points by

$$
(7.73) \quad \omega_{i,j}^{(k+1)} = \left[\frac{1}{1 + \frac{4\Delta t}{\Re h^2}}\right] \left\{ \omega_{i,j}^{(k-1)} - \frac{\Delta t}{h} \left[(u_{i+1,j}^{(k)})(\omega_{i+1,j}^{(k)}) - (u_{i-1,j}^{(k)})(\omega_{i-1,j}^{(k)}) \right. \right.
$$
$$
\left. + (v_{i,j+1}^{(k)})(\omega_{i,j+1}^{(k)}) - (v_{i,j-1}^{(k)})(\omega_{i,j-1}^{(k)}) \right] + \frac{2\Delta t}{\Re h^2} \left[\omega_{i+1,j}^{(k)} + \omega_{i-1,j}^{(k)} \right.
$$
$$
\left.\left. + \omega_{i,j+1}^{(k)} + \omega_{i,j-1}^{(k)} - 2\omega_{i,j}^{(k-1)} \right] \right\} .
$$

Next, generate $\psi_{i,j}^{(k+1)}$ by Method D, using

(7.74) $\quad -4\psi_{i,j}^{(k+1)} + \psi_{i-1,j}^{(k+1)} + \psi_{i,j+1}^{(k+1)} + \psi_{i+1,j}^{(k+1)} + \psi_{i,j+1}^{(k+1)} = -h^2 \omega_{i,j}^{(k+1)}$.

Finally, generate $\omega_{i,j}^{(k+1)}$ at boundary grid points by means of (7.55).

The stability criterion used by Fromm is

(7.75) $\quad\quad\quad \Delta t < \min\left[\frac{\Re}{4} h^2, \frac{h}{u_{max}}\right]$.

7.7 The Method of Pearson

Just as we improved the methods for parabolic problems by proceeding from explicit to implicit methods, Pearson has attempted, essentially, a similar improvement in the method of Fromm by generating $\omega_{i,j}^{(k+1)}$ implicitly from

$$(7.76) \quad \frac{\omega_{i,j}^{(k+1)} - \omega_{i,j}^{(k)}}{\Delta t} = \frac{1}{\mathcal{R}}\left[\frac{1}{2}\Delta_h \omega_{i,j}^{(k)} + \frac{1}{2}\Delta_h \omega_{i,j}^{(k+1)}\right] + \frac{1}{16h^2}\left[\omega_{i,j+1}^{(k+1)}\right.$$

$$\left. + \omega_{i,j+1}^{(k)} - \omega_{i,j-1}^{(k+1)} - \omega_{i,j-1}^{(k)}\right]\left[\psi_{i+1,j}^{(k+1)} + \psi_{i+1,j}^{(k)} - \psi_{i-1,j}^{(k+1)}\right.$$

$$\left. - \psi_{i-1,j}^{(k)}\right] - \frac{1}{16h^2}\left[\omega_{i+1,j}^{(k+1)} + \omega_{i+1,j}^{(k)} - \omega_{i-1,j}^{(k+1)}\right.$$

$$\left. - \omega_{i-1,j}^{(k)}\right]\left[\psi_{i,j+1}^{(k+1)} + \psi_{i,j+1}^{(k)} - \psi_{i,j-1}^{(k+1)} - \psi_{i,j-1}^{(k)}\right],$$

where

$$(7.77) \quad \Delta_h \omega_{i,j} = \frac{-4\omega_{i,j} + \omega_{i+1,j} + \omega_{i,j+1} + \omega_{i-1,j} + \omega_{i,j-1}}{h^2}.$$

It is interesting to note, however, that Pearson has to modify Fromm's method further in that, in order to converge to a steady state solution, he requires smoothing and convergence of a double sequence, as shown in Figure 7.2, at *each* time step.

7.8 Remarks on Three Dimensional Problems

For nonsymmetric, three dimensional Navier-Stokes time dependent problems, one cannot transform to a stream function and vorticity formulation, and one must study the problem directly in terms of velocity components. Such problems are basic, for example, in numerical weather prediction. The methods available, like that of Chorin, are limited by such stability conditions as

(7.78) $$R \Delta t \ll 1$$

and require relatively large amounts of computation time. The method of Chorin is a step ahead method which incorporates, in addition, a predictor-corrector technique. Generally speaking, much work has yet to be done in this vital area.

7.9 Hyperbolic Systems

We turn finally to the study of the motion of a fluid which has the characteristics of a gas. Problems in gas dynamics are usually divided into two categories, those of low speed and those of high speed, because high speed motion is characterized by the emergence of shock waves, which are nonexistent in the low speed case.

Though there are a large number of equations thought to model gas flow at low speeds, the numerical solution of these equations often can be accomplished by the methods already developed. Thus, the steady state model of von Mises can be solved numerically by

HYPERBOLIC SYSTEMS

the methods of Chapter 6 (see, e.g., Greenspan and Jain), while the time varying model of Batchelor can be solved numerically by the methods of Chapter 7 (see, e.g., Schultz). For this reason, we will direct attention only to the case of high speed flow, which will require the development of entirely new numerical methods.

The high speed flow of a compressible, inviscid fluid, like a gas, is usually modeled in terms of hyperbolic systems of first order partial differential equations (Courant and Friedrichs) and it is such systems which will be studied first. For mathematical clarity the discussion will be limited to two equations in the two dependent variables v and w and the two independent variables x and y. Generalization to other systems follows in a natural way.

Consider the problem of finding a pair of functions $v(x,y)$, $w(x,y)$ which satisfy a system of first order partial differential equations of the form

(7.79)
$$a_{11} v_x + a_{12} w_x + b_{11} v_y + b_{12} w_y + d_1 = 0$$
$$a_{21} v_x + a_{22} w_x + b_{21} v_y + b_{22} w_y + d_2 = 0.$$

If $a_{i,j}$, $b_{i,j}$, d_i depend only on x and y, then system (7.79) is said to be linear. If, however, they can depend on x, y, v and w, then the system is said to be quasilinear.

Analysis of system (7.79) is made easier by use of matrix notation as follows. Let

(7.80) $\quad A = \begin{pmatrix} a_{11} & a_{12} \\ a_{21} & a_{22} \end{pmatrix}, \quad B = \begin{pmatrix} b_{11} & b_{12} \\ b_{21} & b_{22} \end{pmatrix}, \quad f = \begin{pmatrix} v \\ w \end{pmatrix}, \quad d = \begin{pmatrix} d_1 \\ d_2 \end{pmatrix}.$

Then (7.79) is equivalent to

(7.81) $\quad\quad\quad\quad\quad\quad Af_x + Bf_y + d = 0.$

System (7.79), or, equivalently, (7.81), is said to be hyperbolic if and only if the determinantal equation

$$|A - \lambda B| = 0$$

has roots λ_1, λ_2 which are real and distinct. Throughout, we will assume that (7.79) is hyperbolic. The solutions of

$$\frac{dx}{dy} = \lambda_1, \quad \frac{dx}{dy} = \lambda_2$$

are called the characteristics of (7.79). Also, note that if T is a two-by-two nonsingular matrix, then (7.81) is equivalent to

(7.82) $\quad\quad\quad\quad\quad TAf_x + TBf_y + Td = 0.$

If, then,

$$C = \begin{pmatrix} \lambda_1 & 0 \\ 0 & \lambda_2 \end{pmatrix}$$

and T is any nonsingular matrix such that

$$TA = CTB,$$

HYPERBOLIC SYSTEMS

then (7.82) is called a <u>normal</u> form of (7.81).

Example 1

Consider the wave equation

(7.83) $$u_{xx} - u_{yy} = 0.$$

If

$$v = u_x, \quad w = u_y,$$

then

(7.84) $$v_x - w_y = 0, \quad v_y - w_x = 0,$$

which is a linear system of type (7.79) which is equivalent to (7.83).

Indeed, (7.84) is of form (7.81) with

$$A = \begin{pmatrix} 1 & 0 \\ 0 & -1 \end{pmatrix}, \quad B = \begin{pmatrix} 0 & -1 \\ 1 & 0 \end{pmatrix}, \quad d = \begin{pmatrix} 0 \\ 0 \end{pmatrix}.$$

Next, since

$$A - \lambda B = \begin{pmatrix} 1 & \lambda \\ -\lambda & -1 \end{pmatrix},$$

it follows that

$$|A - \lambda B| = \begin{vmatrix} 1 & \lambda \\ -\lambda & -1 \end{vmatrix} = (-1) + \lambda^2 = 0$$

implies

(7.85) $$\lambda_1 = 1, \quad \lambda_2 = -1.$$

Thus, system (7.84) is hyperbolic, which is consistent with (7.83) being called a hyperbolic second order equation. The characteristics of the system are the solutions of

$$\frac{dx}{dy} = 1, \quad \frac{dx}{dy} = -1,$$

or

$$x - y = c_1, \quad x + y = c_2.$$

Finally, to get a normal form of (7.84), let

$$T = \begin{pmatrix} \alpha & \beta \\ \gamma & \delta \end{pmatrix}.$$

Then we wish to have

$$TA = CTB.$$

But

$$TA = \begin{pmatrix} \alpha & -\beta \\ \gamma & -\delta \end{pmatrix}, \quad TB = \begin{pmatrix} \beta & -\alpha \\ \delta & -\gamma \end{pmatrix}$$

$$CTB = \begin{pmatrix} 1 & 0 \\ 0 & -1 \end{pmatrix} \begin{pmatrix} \beta & -\alpha \\ \delta & -\gamma \end{pmatrix} = \begin{pmatrix} \beta & -\alpha \\ -\delta & \gamma \end{pmatrix},$$

so that one wishes to have

$$\begin{pmatrix} \alpha & -\beta \\ \gamma & -\delta \end{pmatrix} = \begin{pmatrix} \beta & -\alpha \\ -\delta & \gamma \end{pmatrix},$$

HYPERBOLIC SYSTEMS

or, equivalently

$$\alpha = \beta$$
$$-\beta = -\alpha$$
$$\gamma = -\delta$$
$$-\delta = \gamma,$$

one solution of which is

$$\alpha = \beta = 1, \quad \gamma = 1, \quad \delta = -1.$$

Thus, one possible choice of T is

$$T = \begin{pmatrix} 1 & 1 \\ 1 & -1 \end{pmatrix}.$$

But then

$$TA = \begin{pmatrix} 1 & 1 \\ 1 & -1 \end{pmatrix}\begin{pmatrix} 1 & 0 \\ 0 & -1 \end{pmatrix} = \begin{pmatrix} 1 & -1 \\ 1 & 1 \end{pmatrix}$$

$$TB = \begin{pmatrix} 1 & 1 \\ 1 & -1 \end{pmatrix}\begin{pmatrix} 0 & -1 \\ 1 & 0 \end{pmatrix} = \begin{pmatrix} 1 & -1 \\ -1 & -1 \end{pmatrix}$$

and system (7.82) is

$$\begin{pmatrix} 1 & -1 \\ 1 & 1 \end{pmatrix}\begin{pmatrix} v_x \\ w_x \end{pmatrix} + \begin{pmatrix} 1 & -1 \\ -1 & -1 \end{pmatrix}\begin{pmatrix} v_y \\ w_y \end{pmatrix} = 0$$

or, equivalently,

(7.86)
$$v_x - w_x + v_y - w_y = 0$$
$$v_x + w_x - v_y - w_y = 0$$

which is a normal form equivalent of (7.84).

Example 2

The one dimensional isentropic flow equations are

(7.87)
$$\rho_t + u\rho_x + \rho u_x = 0$$
$$\rho(u_t + uu_x) + c^2 \rho_x = 0,$$

where u is velocity, ρ is density, and $c = c(\rho)$ is the speed of sound. With the change of variables

$$v = u, \quad w = \rho, \quad y = t,$$

system (7.87) can be rewritten as

(7.87')
$$w_y + vw_x + wv_x = 0$$
$$wv_y + wvv_x + c^2 w_x = 0.$$

This is a quasilinear system of form (7.79) with

$$A = \begin{pmatrix} w & v \\ wv & c^2 \end{pmatrix}, \quad B = \begin{pmatrix} 0 & 1 \\ w & 0 \end{pmatrix}, \quad d = \begin{pmatrix} 0 \\ 0 \end{pmatrix}.$$

Hence,

$$A - \lambda B = \begin{pmatrix} w & v-\lambda \\ wv-\lambda w & c^2 \end{pmatrix},$$

so that

(7.88)
$$|A - \lambda B| = wc^2 - w(v-\lambda)^2 = 0.$$

HYPERBOLIC SYSTEMS

Since $w = \rho$ is the density, assume $w \neq 0$, so that (7.88) implies

$$c^2 - (v - \lambda)^2 = 0,$$

or

$$\lambda = v \pm c.$$

Since c is the speed of sound, assume $c \neq 0$, so that

$$\lambda_1 = v + c, \quad \lambda_2 = v - c.$$

The characteristics of system (7.87') are the solutions of

$$\frac{dx}{dy} = v + c, \quad \frac{dx}{dy} = v - c,$$

which <u>cannot</u> be solved because v is unknown. To get a normal form of (7.87'), let

$$T = \begin{pmatrix} \alpha & \beta \\ \gamma & \delta \end{pmatrix}.$$

Then

$$TA = \begin{pmatrix} \alpha & \beta \\ \gamma & \delta \end{pmatrix} \begin{pmatrix} w & v \\ wv & c^2 \end{pmatrix} = \begin{pmatrix} \alpha w + \beta wv & \alpha v + \beta c^2 \\ \gamma w + \delta wv & \gamma v + \delta c^2 \end{pmatrix}$$

$$TB = \begin{pmatrix} \alpha & \beta \\ \gamma & \delta \end{pmatrix} \begin{pmatrix} 0 & 1 \\ w & 0 \end{pmatrix} = \begin{pmatrix} \beta w & \alpha \\ \delta w & \gamma \end{pmatrix}$$

$$CTB = \begin{pmatrix} v+c & 0 \\ 0 & v-c \end{pmatrix} \begin{pmatrix} \beta w & \alpha \\ \delta w & \gamma \end{pmatrix} = \begin{pmatrix} v\beta w + c\beta w & v\alpha + c\alpha \\ v\delta w - c\delta w & v\gamma - c\gamma \end{pmatrix}.$$

Thus,

$$TA = CTB$$

which implies

$$\begin{pmatrix} \alpha w + \beta wv & \alpha v + \beta c^2 \\ \gamma w + \delta wv & \gamma v + \delta c^2 \end{pmatrix} = \begin{pmatrix} v\beta w + c\beta w & v\alpha + c\alpha \\ v\delta w - c\delta w & v\gamma - c\gamma \end{pmatrix},$$

or, since $c \neq 0$, $w \neq 0$,

$$\alpha = c\beta$$
$$\alpha = c\beta$$
$$\gamma = -c\delta$$
$$\gamma = -c\delta.$$

A solution of the latter system is $\beta = 1$, $\alpha = c$, $\delta = 1$, $\gamma = -c$, so that one can choose

$$T = \begin{pmatrix} c & 1 \\ -c & 1 \end{pmatrix}.$$

Then,

$$TA = \begin{pmatrix} cw + vw & cv + c^2 \\ -cw + vw & -cv + c^2 \end{pmatrix}$$

$$TB = \begin{pmatrix} w & c \\ w & -c \end{pmatrix},$$

and a normal form is

$$\begin{pmatrix} cw + vw & cv + c^2 \\ -cw + vw & -cv + c^2 \end{pmatrix} \begin{pmatrix} v_x \\ w_x \end{pmatrix} + \begin{pmatrix} w & c \\ w & -c \end{pmatrix} \begin{pmatrix} v_y \\ w_y \end{pmatrix} = 0,$$

or, equivalently,

INITIAL VALUE PROBLEMS

(7.89)
$$(cw + vw)v_x + (cv + c^2)w_x + wv_y + cw_y = 0$$
$$(-cw + vw)v_x + (-cv + c^2)w_x + wv_y - cw_y = 0.$$

7.10 Initial Value Problems

It can be shown that any hyperbolic system (7.79) has a normal form, so that in considering initial value problems for hyperbolic systems, we assume that the system is already in normal form. We allow the normal form to be quasilinear. In formulating initial value problems, we assume v and w are given on some curve, which, for simplicity, will be taken as the X-axis. It will be assumed also that the X-axis is not a characteristic of the given system. We will wish to find solutions v, w of the system for $y > 0$ which take on the given initial values on the X-axis. More precisely, an initial value problem is defined as follows. Given

(7.90)
$$\begin{cases} v(x,0) = g(x), & \text{on X axis} \\ w(x,0) = h(x), & \text{on X axis} \\ C = \begin{pmatrix} c_1 & 0 \\ 0 & c_2 \end{pmatrix}; & c_1, c_2 \text{ real and unequal} \\ A = \begin{pmatrix} a_{11} & a_{12} \\ a_{21} & a_{22} \end{pmatrix}; & |A| \neq 0, \end{cases}$$

then, for $y > 0$, find $v(x,y), w(x,y)$ which satisfy initial conditions (7.90), and which for $y > 0$ are solutions of

$$CAf_x + Af_y + d = 0,$$

or, equivalently, of

(7.91) $$\begin{pmatrix} c_1 a_{11} & c_1 a_{12} \\ c_2 a_{21} & c_2 a_{22} \end{pmatrix} \begin{pmatrix} v_x \\ w_x \end{pmatrix} + \begin{pmatrix} a_{11} & a_{12} \\ a_{21} & a_{22} \end{pmatrix} \begin{pmatrix} v_y \\ w_y \end{pmatrix} + \begin{pmatrix} d_1 \\ d_2 \end{pmatrix} = 0,$$

or, equivalently, of

(7.92)
$$c_1 a_{11} v_x + c_1 a_{12} w_x + a_{11} v_y + a_{12} w_y + d_1 = 0$$
$$c_2 a_{21} v_x + c_2 a_{22} w_x + a_{21} v_y + a_{22} w_y + d_2 = 0.$$

Existence and uniqueness of initial value problem (7.90)–(7.92) is known only in the small (Courant and Hilbert), and since no known analytical method is available for solving it, we turn next to a very general, useful scheme for approximating a solution.

7.11 The Method of Courant, Isaacson and Rees

Of basic importance in the method to be developed are the difference approximations of (7.91) and (7.92). For the Method of Courant, Isaacson and Rees, these are given as follows. Fix $\Delta x = h$, $\Delta y = k$ and number the four points (x,y), $(x+h,y)$, $(x,y+k)$, $(x-h,y)$ with 0, 1, 2, 3, as shown in Figure 7.20. Then approximate (7.91) by

(7.93) $$c_1(0) a_{11}(0) \alpha + c_1(0) a_{12}(0) \beta + a_{11}(0) \left(\frac{v_2 - v_0}{k} \right)$$
$$+ a_{12}(0) \left(\frac{w_2 - w_0}{k} \right) + d_1(0) = 0$$

METHOD OF COURANT, ISAACSON AND REES

and

$$(7.94) \quad c_2(0) a_{21}(0) \gamma + c_2(0) a_{22}(0) \delta + a_{21}(0) \left(\frac{v_2 - v_0}{k}\right)$$

$$+ a_{22}(0) \left(\frac{w_2 - w_0}{k}\right) + d_2(0) = 0,$$

where

$$\alpha = \frac{v_0 - v_3}{h}, \quad \beta = \frac{w_0 - w_3}{h}, \quad \text{if } c_1(0) \geq 0$$

$$\alpha = \frac{v_1 - v_0}{h}, \quad \beta = \frac{w_1 - w_0}{h}, \quad \text{if } c_1(0) < 0$$

$$\gamma = \frac{v_0 - v_3}{h}, \quad \delta = \frac{w_0 - w_3}{h}, \quad \text{if } c_2(0) \geq 0$$

$$\gamma = \frac{v_1 - v_0}{h}, \quad \delta = \frac{w_1 - w_0}{h}, \quad \text{if } c_2(0) < 0.$$

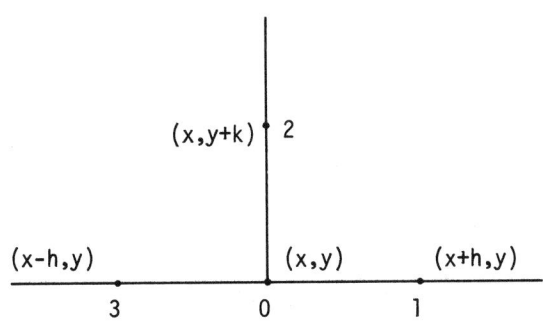

Figure 7.20

The algorithm can now be formulated as follows.

<u>Step 1</u> For $\Delta x = h$, $\Delta y = k$, construct grid points in the upper half plane.

<u>Step 2</u> Beginning on the X-axis, generate approximations for v and w from three known values at (x,y), $(x+h,y)$, $(x-h,y)$, as shown in Figure 7.20, at the point $(x,y+k)$ by writing down (7.93) and (7.94) and then solving these two equations for v_2, w_2.

<u>Step 3</u> Continue from row k to row k+1, k = 1,2,3,..., as indicated in Step 2.

For linear systems, the stability condition for the above method is

(7.95) $$\frac{h}{k} \geq \max(|c_1|, |c_2|),$$

which is to be valid over the entire region of interest. For quasilinear systems, (7.95) may be of little, or no, value.

<u>Example</u>

Consider the isentropic flow equations (7.87) in normal form (7.89) with c = 1100, that is,

$$(1100 + v)wv_x + (1100)(v + 1100)w_x + wv_y + 1100 w_y = 0$$
$$(-1100 + v)wv_x + (1100)(1100 - v)w_x + wv_y - 1100 w_y = 0.$$

METHOD OF COURANT, ISAACSON AND REES

Let

$$v(x,0) = 0, \quad w(x,0) = x.$$

Let us approximate $v(1,.1)$, $w(1,.1)$.

Since the system is in normal form,

$$a_{11} = w, \quad a_{12} = 1100, \quad a_{21} = w, \quad a_{22} = -1100$$

$$c_1 = 1100 + v, \quad c_2 = v - 1100, \quad d_1 = d_2 = 0.$$

Choose $h = 1$, $k = .1$, and number the points $(1,0)$, $(2,0)$, $(1,.1)$, $(0,0)$ by 0, 1, 2, 3, as shown in Figure 7.21.

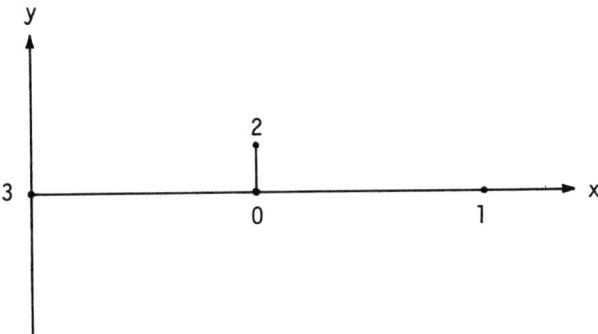

Figure 7.21

Then,

$$v_0 = v(1,0) = 0 \qquad w_0 = w(1,0) = 1$$

$$v_1 = v(2,0) = 0 \qquad w_1 = w(2,0) = 2$$

$$v_3 = v(0,0) = 0 \qquad w_3 = w(0,0) = 0.$$

254   FLUID PROBLEMS

Hence,

$$a_{11}(0) = w_0 = 1 \qquad a_{21}(0) = w_0 = 1$$
$$c_1(0) = 1100 + v_0 = 1100 \qquad c_2(0) = v_0 - 1100 = -1100$$
$$a_{12}(0) = 1100 \qquad a_{22}(0) = -1100 \; .$$
$$d_1(0) = d_2(0) = 0$$

Thus, (7.93) and (7.94) become

$$(1100 + v_0)w_0 \left(\frac{v_0 - v_3}{1}\right) + 1100(v_0 + 1100)\left(\frac{w_0 - w_3}{1}\right)$$
$$+ w_0 \left(\frac{v_2 - v_0}{1/10}\right) + 1100\left(\frac{w_2 - w_0}{1/10}\right) = 0$$

$$(v_0 - 1100)w_0 \left(\frac{v_1 - v_0}{1}\right) + (1100)(1100 - v_0)\left(\frac{w_1 - w_0}{1}\right)$$
$$+ w_0 \left(\frac{v_2 - v_0}{1/10}\right) - 1100\left(\frac{w_2 - w_0}{1/10}\right) = 0 \; ,$$

or, equivalently,

$$v_2 + 1100\, w_2 = -(1100)(109)$$
$$v_2 - 1100\, w_2 = -(1100)(111)$$

the solution of which is the desired approximation.

7.12 <u>The Lax-Wendroff Method</u>

In gas dynamical problems, one often encounters a hyperbolic system of equations in the special form

LAX-WENDROFF METHOD

$$\frac{\partial v}{\partial t} + \frac{\partial}{\partial x}[f_1(v,w)] = 0$$

$$\frac{\partial w}{\partial t} + \frac{\partial}{\partial x}[f_2(v,w)] = 0.$$

Such a system is said to be in conservative form because f_1 and f_2 often represent such conserved quantities as energy, mass, or momentum. For such systems the Lax-Wendorff method is of particular value and can be outlined as follows.

If one has determined v and w on the line $t = t^*$ of a rectangular grid with grid sizes Δx and Δt, then one proceeds to grid points with $t = t^* + \Delta t$ in two steps. First one determines v and w at centers of rectangle mesh areas by

$$v(x + \tfrac{\Delta x}{2}, t^* + \tfrac{\Delta t}{2}) = \tfrac{1}{2}[v(x+\Delta x, t^*) + v(x, t^*)]$$

$$- \tfrac{\Delta t}{2\Delta x}\{f_1(v(x+\Delta x, t^*), w(x+\Delta x, t^*)) - f_1(v(x, t^*), w(x, t^*))\}$$

$$w(x + \tfrac{\Delta x}{2}, t^* + \tfrac{\Delta t}{2}) = \tfrac{1}{2}[w(x+\Delta x, t^*) + w(x, t^*)]$$

$$- \tfrac{\Delta t}{2\Delta x}\{f_2(v(x+\Delta x, t^*), w(x+\Delta x, t^*)) - f_2(v(x, t^*), w(x, t^*))\}.$$

Then, determine v and w at mesh points of the form $(x, t^* + \Delta t)$ from

$$v(x, t^* + \Delta t) = v(x, t^*) - \tfrac{\Delta t}{\Delta x}\{f_1(v(x+\tfrac{\Delta x}{2}, t^* + \tfrac{\Delta t}{2}), w(x+\tfrac{\Delta x}{2}, t^* + \tfrac{\Delta t}{2}))$$

$$- f_1(v(x - \tfrac{\Delta x}{2}, t^* + \tfrac{\Delta t}{2}), w(x - \tfrac{\Delta x}{2}, t^* + \tfrac{\Delta t}{2}))\}$$

$$w(x,t^* + \Delta t) = w(x,t^*) - \frac{\Delta t}{\Delta x} \{f_2(v(x+\frac{\Delta x}{2}, t^* + \frac{\Delta t}{2}), w(x+\frac{\Delta x}{2}, t^* + \frac{\Delta t}{2})$$

$$- f_2(v(x-\frac{\Delta x}{2}, t^* + \frac{\Delta t}{2}), w(x-\frac{\Delta x}{2}, t^* + \frac{\Delta t}{2}))\} .$$

For a relatively concise discussion of stability conditions for the Lax-Wendroff method and of its application to the construction of shock waves, see Ames.

7.13 Other Methods

With regard to other available methods for fluid problems, the most notable is the particle-in-cell method (see Amsden and the references contained therein). This is an expensive but valuable method which has been applied with apparent success to very broad categories of fluid dynamics problems. With regard to boundary layer calculations, the method of Spaulding is both interesting and promising.

Exercises

1. Derive formulas (7.13)–(7.15).

2. Prove that application of (7.30)–(7.33) on the inner boundary preserves diagonal dominance.

3. Let $A(0,0)$, $B(1,0)$, $C(1,2)$, $D(0,2)$ be a rectangular cavity. Consider the two dimensional, steady state, Navier–Stokes equations (7.1)–(7.2) on R and boundary conditions (7.3)–(7.6) on S. Using $h = 0.1$, generate numerical solutions for the three cases $\mathcal{R} = 1$, 100, and 1000, and graph the resulting flows. Use smoothing in all three cases.

4. Generate the numerical solution of the initial value problem (7.41)–(7.46) for each of $\mathcal{R} = 1$, 10, and 100 by the method of Section 7.5.

5. Generate the numerical solution of initial value problem (7.41)–(7.46) for each of $\mathcal{R} = 1$, 10, and 100 by the method of Fromm and compare your results with those of Exercise 4.

6. Generate the numerical solution of the initial value problem (7.41)–(7.46) for each of $\mathcal{R} = 1$, 10, and 100 by the method of Pearson and compare your results with those of Exercises 4 and 5.

7. Determine which of the following systems are hyperbolic. For those which are, find the characteristics and a normal form.

(a) $v_x + w_x + v_y - w_y = 0$
$v_x + w_x - v_y - w_y = 0$

(b) $v_x - w_x + v_y - w_y = 0$
$v_x - w_x - v_y - w_y = 0$

(c) $v_x - w_x + v_y - w_y = 0$
$v_x + w_x - v_y - w_y = 0$

(d) $v_x + w_y = 0$
$v_y + w_x = 0$

8. Find a normal form, different from (7.89), for the one dimensional, isentropic flow equations (7.87).

9. Consider the isentropic flow equations (7.89) with $c = 1100$. Let $v(x,0) = x$, $w(x,0) = x^2$. Approximate $v(1,0.1)$ and $w(1,0.1)$ by the method of Section 7.11 for each of the following cases.

(a) $h = 1$, $k = 0.1$
(b) $h = 1$, $k = 0.05$
(c) $h = 0.5$, $k = 0.1$
(d) $h = 0.1$, $k = 0.01$.

CHAPTER VIII

DISCRETE MODEL THEORY

8.1 Introduction

From the mathematical point of view, it is somewhat less than satisfying to realize that, for realistic nonlinear models, rarely are we able to show that the numerical solution generated by a finite difference technique converges to an analytical solution as the grid size converges to zero. A natural way out of this plight is simply to take the difference equation one uses, rather than the given differential equation, as the dynamical equation. The selection of a dynamical difference equation ab ovo, without any consideration of a differential equation, is the essential substance of discrete model theory.

In this chapter we will illustrate ideas and methods of discrete model theory by developing basic discrete, plane Newtonian dynamics, thereby constructing a physical theory which will be entirely arithmetic, and therefore completely compatible with digital computer capabilities.

8.2 Particles, Time, and Motion

From a purely mathematical point of view, one can consider the terms particle, time, and motion as undefined and can then proceed to define other concepts in terms of these. (The reader unfamiliar with the role of undefined terms in a mathematical science should, at this point, read the

Appendix). Nevertheless, physically, it is desirable to have some intuition about these rudiments and in this section we will try to develop such intuition.

A particle of a given solid will be considered to be a small spherical portion of the solid. A plane particle, which is the only kind with which we will deal, will be any great circle section of a particle. The centroid of a plane particle is defined to be the center of the associated great circle. By the position of a particle we will mean the position of its centroid. The mass of a particle is defined in the usual Newtonian way, and all plane figures are to be considered as compositions of particles.

With regard to the concept of motion, let Δx and Δt be positive constants. On an X-axis, mark off $X_k = k\Delta x$, $k = 0,1,\ldots,m$, and, on a T-axis, mark off $t_j = j\Delta t$, $j = 0,1,\ldots,n$. For illustrative purposes, suppose that a particle P is located at X_0 when $t = t_0$, at X_3 when $t = t_1$, at X_{10} when $t = t_2$ and at X_5 when $t = t_3$. The motion of P from X_0 to X_5 is merely P's being at X_0, X_3, X_{10}, X_5 at the respective times t_0, t_1, t_2, t_3. Thus, the motion of P from X_0 to X_5 is considered to be a sequence of four "stills". This concept of motion is physiologically acceptable and is realized in motion pictures, where the viewer observes motion from a finite sequence of projected stills.

8.3 Velocity and Acceleration

Consider, now, the basic concepts of velocity and acceleration and, for simplicity, let us continue to confine our attention to a particle moving in a fixed direction.

For $\Delta t > 0$, let $t_k = k\Delta t$, $k = 0, 1, \ldots, n-1, n$. At time t_k, let a particle which is in motion in a fixed X direction have its center at x_k. We wish to define the velocity v_k and acceleration a_k of the particle at each time t_k. Consider, then, first the interval from t_0 to t_1. Suppose, in addition to x_0 and x_1, one knows v_0, as would be the case in a falling body problem when the particle's motion begins from a position of rest, so that one could assume $v_0 = 0$. Let us try to define $v_1 = v(t_1)$ in a fashion that will use all the given data. This can be accomplished, for example, by defining v_1 implicitly by the smoothing formula

$$(8.1) \qquad \frac{x_1 - x_0}{\Delta t} = \frac{v_0 + v_1}{2},$$

which then motivates our general definition

$$(8.2) \qquad \frac{x_k - x_{k-1}}{\Delta t} = \frac{v_{k-1} + v_k}{2}, \quad k = 1, 2, \ldots, n,$$

for velocity v_k. Of course, an equivalent form of (8.2) is

$$(8.3) \qquad v_k = v_{k-1} + \frac{2}{\Delta t}(x_k - x_{k-1}), \quad k = 1, 2, \ldots, n.$$

With regard to acceleration $a_k = a(t_k)$, $k = 1, \ldots, n$, one rarely knows a_0 without knowing the force in action, so that it is reasonable to define a_0 by the forward difference

(8.4) $$a_0 = \frac{v_1 - v_0}{\Delta t},$$

from which we are motivated to define a_k, in general, by

(8.5) $$a_{k-1} = \frac{v_k - v_{k-1}}{\Delta t}, \quad k = 1, 2, \ldots, n+1.$$

8.4 The Law of Motion

To determine the motion of a particle acted upon by a given force, it is usual to relate force and acceleration by a dynamical equation. For this purpose, let a particle of mass m be in motion on an X axis and be located at x_k at time $t_k = k\Delta t$, $k = 0, 1, \ldots, n$. At time t_k, let the particle be acted upon by a force $F = F(t_k, x_k, v_k)$. Then the motion of the particle is assumed to be governed by a discrete Newton's equation:

(8.6) $$m \cdot a(t_k) = F(t_k, x_k, v_k); \quad k = 0, 1, \ldots, n.$$

The values of F can be given either in tabular form from experimental data, or in the form of a mathematical expression.

8.5 Damped Motion in a Nonlinear Force Field

Before proceeding to more theoretical questions, let us show how easily the formulation given thus far can be implemented. For this purpose attention will be directed to the study of damped, oscillatory motion in a nonlinear force field.

DAMPED, NONLINEAR MOTION

Consider a particle P of unit mass which is constrained to move with its center C on an X axis. A displacement of the particle such that the directed distance OC is x_i is, for illustrative purposes, assumed to be opposed by a field force of magnitude $\sin x_i$ and by a viscous damping force of magnitude αv_i, where α is a positive constant. Such a set of interacting forces is typical in the analysis of the motion of a pendulum, as discussed in Section 2.5. Then the equation of motion (8.6) takes the particular form

(8.7) $$a_k = -\alpha v_k - \sin x_k, \quad k = 0,1,2,\ldots,n.$$

But, from (8.2) and (8.5),

(8.8) $$v_k = v_{k-1} + a_{k-1}\Delta t, \quad k = 1,2,\ldots,n$$

and

(8.9) $$x_k = x_{k-1} + \frac{\Delta t}{2}(v_k + v_{k-1}), \quad k = 1,2,\ldots,n,$$

so that the motion of C can be generated recursively as follows. Fix x_0 and v_0, that is, an initial position and an initial velocity, respectively. Generate a_0 from (8.7), then v_1 from (8.8), and finally position x_1 from (8.9). Next, using x_1 and v_1, calculate a_1 from (8.7), then v_2 from (8.8) and finally x_2 from (8.9). In the indicated fashion, from x_{k-1} and v_{k-1}, generate a_{k-1} from (8.7), then v_k from (8.8), and finally x_k from (8.9).

Since, for any x_{k-1} and v_{k-1}, (8.7)-(8.9) imply that a_{k-1}, v_k and x_k exist and are unique, it follows immediately from the above discussion that the motion of C is uniquely defined once initial conditions x_0 and v_0 are given, or, in more general terminology, the solution of an initial value problem for (8.7) exists and is unique.

As an illustrative example, the solution of (8.7)-(8.9) with the parameter values $\alpha = 0.3, \Delta t = 0.01, x_0 = \pi/4, v_0 = 0, n = 15000$, was generated on the UNIVAC 1108 in under 30 seconds, and the results were completely analogous to those of Section 2.5.

8.6 Conservation of Energy

Let us show now that, unlike the difference approximations developed in Chapter II, the present formulation is, in fact, energy conserving. For this purpose, the force to be considered will be gravity.

Let $t_k = k\Delta t, k = 0, 1, \ldots, n$. At time t_i, let a particle be located at point (x_i, y_i), which is on the straight line through $A(x_0, y_0)$ and $B(x_n, y_n)$, one possible arrangement of which is shown in Figure 8.1. Let $F(t_i)$ represent the component in the direction $\vec{AB}$ of a force $\vec{F}$ applied to the particle. Then the work W done by $\vec{F}$ in moving the particle in the fixed direction from A to B is defined to be

CONSERVATION OF ENERGY

(8.10) $$W = \sum_{i=1}^{n} F(t_{i-1}) \Delta s_i ,$$

where

(8.11) $$\Delta s_i = s_i - s_{i-1}$$

is the <u>directed</u> distance from (x_{i-1}, y_{i-1}) to (x_i, y_i).

Now let a_i and v_i represent the acceleration and velocity, respectively, at (x_i, y_i) in the direction $\vec{AB}$. Then, from (8.6) and (8.10) it follows that

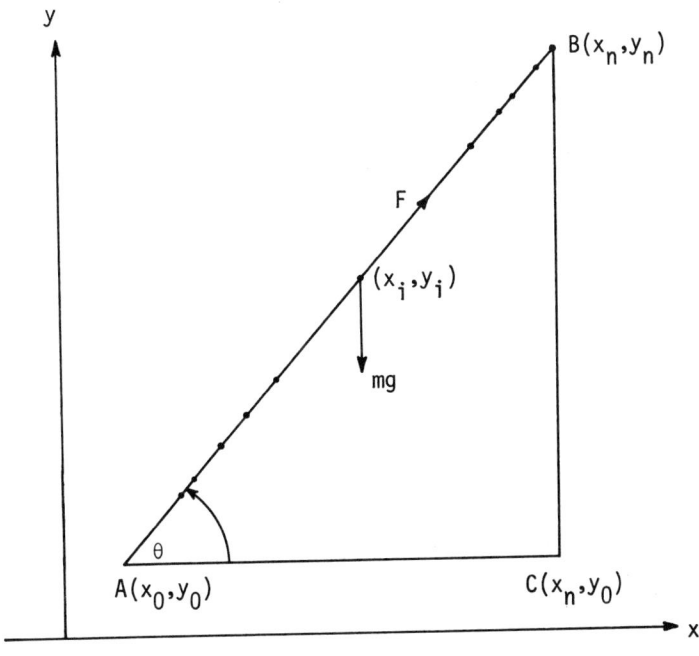

Figure 8.1

$$W = \sum_{i=1}^{n} F(t_{i-1}) \Delta s_i$$

$$= m \sum_{i=1}^{n} a_{i-1} \Delta s_i$$

$$= m \sum_{i=1}^{n} \left[\left(\frac{v_i - v_{i-1}}{\Delta t} \right)(s_i - s_{i-1}) \right]$$

$$= m \sum_{i=1}^{n} \left[(v_i - v_{i-1}) \left(\frac{v_i + v_{i-1}}{2} \right) \right]$$

$$= \frac{m}{2} \sum_{i=1}^{n} [v_i^2 - v_{i-1}^2].$$

Thus,

(8.12) $$W = \frac{mv_n^2}{2} - \frac{mv_0^2}{2}.$$

The quantity

(8.13) $$K_i = \frac{1}{2} mv_i^2$$

is defined to be the kinetic energy of the particle at time t_i, and from (8.11) and (8.12), one has

(8.14) $$W = K_n - K_0.$$

Neglecting friction, the force necessary to move a particle of mass m only along the vertical component of $\vec{AB}$ must be equal to the weight mg of the particle. Hence, the amount of work done along the vertical component of the motion is

NONLINEAR STRING VIBRATIONS

(8.15) $$W = mg(y_n - y_0),$$

so that

(8.16) $$W = mg\, y_n - mg\, y_0.$$

The quantity

(8.17) $$V_i = -mg\, y_i$$

is called the gravitational potential energy of the particle at the point (x_i, y_i), so that, from (8.14) and (8.15),

(8.18) $$W = -V_n + V_0.$$

Finally, from (8.13) and (8.16), one has

(8.19) $$K_n + V_n = K_0 + V_0, \quad n = 0,1,2,3,\ldots,$$

which is called the **principle** of **conservation** of **energy**.

Thus, the discrete Newton's equation (8.16) and our particular definitions of velocity, acceleration and work have yielded a fundamental conservation principle with regard to gravity in a completely arithmetic setting.

8.7 Nonlinear String Vibrations.

Consider next the motion of a **system** of particles, each of which is moving in a fixed direction. Such a system can be realized nicely in the study of the vibrations of a string, to which this section is directed.

A discrete string is one which is composed of a finite number of particles. It will be treated mathematically as an ordered set of $m+2$ circular, homogeneous particles P_k, $k = 0,1,2,\ldots,m,m+1$, as shown typically in Figure 8.2. Our problem will be that of describing the return of a discrete string to a position of equilibrium from an arbitrary position of tension. The resulting motion can be considered as an approximation to that of a real string, the improvement of which is dependent largely upon one's computer capability. It will be assumed throughout that P_0 and P_{m+1} are fixed, that $P_1, P_2, \ldots, P_m$ are free to move, but in the vertical direction only, and that

(8.20) $$x_0 = y_0 = y_{m+1} = 0.$$

Let $x_0 < x_1 < x_2 < \cdots < x_m < x_{m+1}$ and $x_i - x_{i-1} = \Delta x$, $i = 1,2,\ldots,m+1$. At time t_k, $k = 0,1,\ldots,n$, measured in seconds, let P_j be a typical particle in motion, with its y components of velocity and acceleration denoted by $v_{j,k}$ and $a_{j,k}$, respectively. In order to incorporate the dependence of the centers of P_{j-1}, P_j and P_{j+1} on time, let the respective centers of these particles at time t_k be $(x_{j-1}, y_{j-1,k})$, $(x_j, y_{j,k})$, $(x_{j+1}, y_{j+1,k})$, where each coordinate is measured in feet.

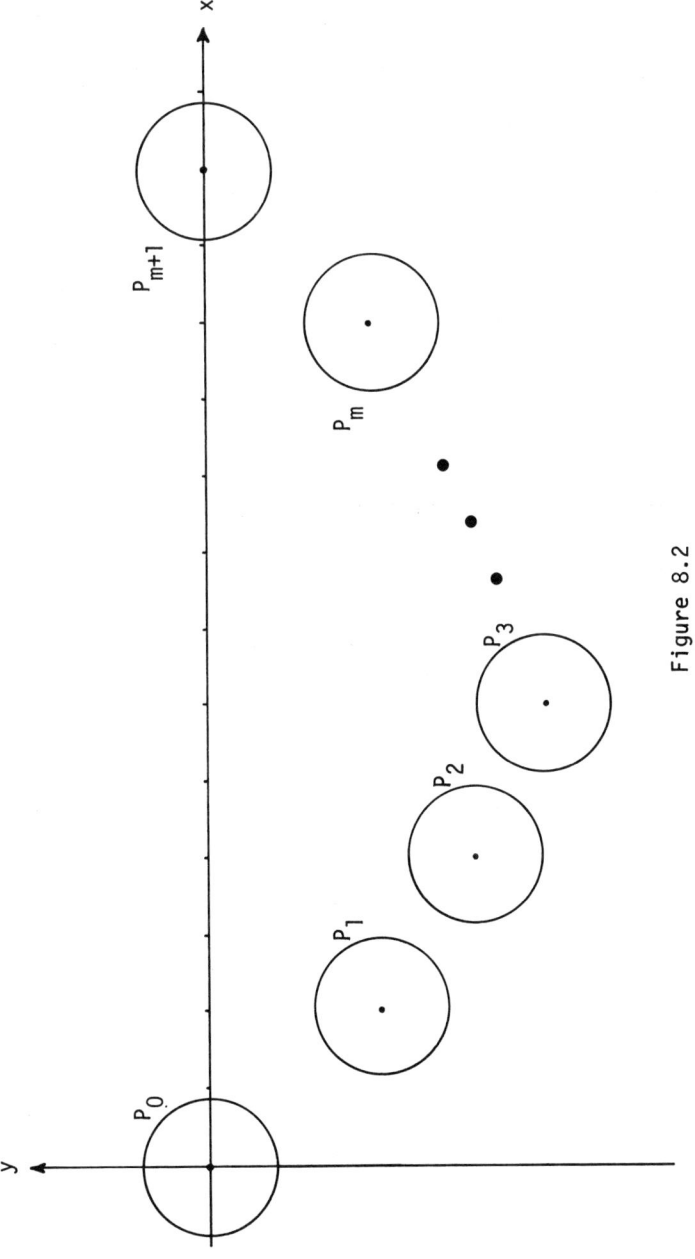

Figure 8.2

In studying the motion of P_j, we will take into account only tensile, viscous, and gravitational forces. For this purpose, let T_1 be the tensile force between P_{j-1} and P_j, let T_2 be the tensile force between P_j and P_{j+1}, and let the viscosity vary with the speed of the particle. Then (8.6) takes the particular form

$$(8.21) \quad |T_2| \frac{(y_{j+1,k} - y_{j,k})}{[(\Delta x)^2 + (y_{j+1,k} - y_{j,k})^2]^{1/2}} - |T_1| \frac{(y_{j,k} - y_{j-1,k})}{[(\Delta x)^2 + (y_{j,k} - y_{j-1,k})^2]^{1/2}}$$

$$- \alpha v_{j,k} - \bar{m} g = \bar{m} a_{j,k}; \quad k = 0,1,2,3,\ldots,$$

where $g \geq 0$, $\alpha \geq 0$, and $\bar{m}$ is the mass of P_j. Thus, from (8.2), (8.5), and (8.21), one has

$$(8.22) \quad a_{j,k} = \frac{|T_2|}{\bar{m}} \frac{(y_{j+1,k} - y_{j,k})}{[(\Delta x)^2 + (y_{j+1,k} - y_{j,k})^2]^{1/2}}$$

$$- \frac{|T_1|}{\bar{m}} \frac{(y_{j,k} - y_{j-1,k})}{[(\Delta x)^2 + (y_{j,k} - y_{j-1,k})^2]^{1/2}}$$

$$- \frac{\alpha v_{j,k}}{\bar{m}} - g,$$

$$(8.23) \quad v_{j,k+1} = v_{j,k} + a_{j,k} \Delta t$$

$$(8.24) \quad y_{j,k+1} = y_{j,k} + \frac{\Delta t}{2} (v_{j,k+1} + v_{j,k}),$$

for $j = 1,2,\ldots,m$ and $k = 0,1,2,\ldots,n$. Actual calculation with (8.22)-(8.24) is completely analogous to that with (8.7)-(8.9) except

NONLINEAR STRING VIBRATIONS

that in the present case one has m particles, instead of a single particle, with which to deal at each time step.

A large number of examples using (8.22)-(8.24) were run at the University of Wisconsin Computing Center and we will describe next a typical such example. The output is given graphically with 100 additional points interpolated linearly between each pair of consecutive particles.

Example

Consider a twenty-one point string with $x_i = \frac{i}{10}$, $i = 0,1,2,\ldots,20$, with T_1 and T_2 defined by

$$(8.25) \quad T_1 = T_0 \left[1 + \left| \frac{y_{i,k-1} - y_{i-1,k-1}}{\Delta x} \right| + \frac{\varepsilon}{2} \left(\frac{y_{i,k-1} - y_{i-1,k-1}}{\Delta x} \right)^2 \right]$$

$$(8.26) \quad T_2 = T_0 \left[1 + \left| \frac{y_{i+1,k-1} - y_{i,k-1}}{\Delta x} \right| + \frac{\varepsilon}{2} \left(\frac{y_{i+1,k-1} - y_{i,k-1}}{\Delta x} \right)^2 \right];$$

and with $\alpha = 0.15$, $\bar{m} = 0.05$, $T_0 = 12.5$, $\Delta t = 0.00025$, $\Delta x = 0.1$, $m = 19$, $g = 0$, $\varepsilon = 0.01$. The string is placed in a position of tension by setting the second particle at $(0,1)$ and by aligning all other particles as shown at $t = 0.00$ in Figure 8.3. The first 0.75 seconds of wave motion are shown typically in Figure 8.3. The development of small trailing waves is readily apparent from the figure.

For a variety of other examples which include more particles and other tension laws, see Greenspan (11). For a discrete formulation of

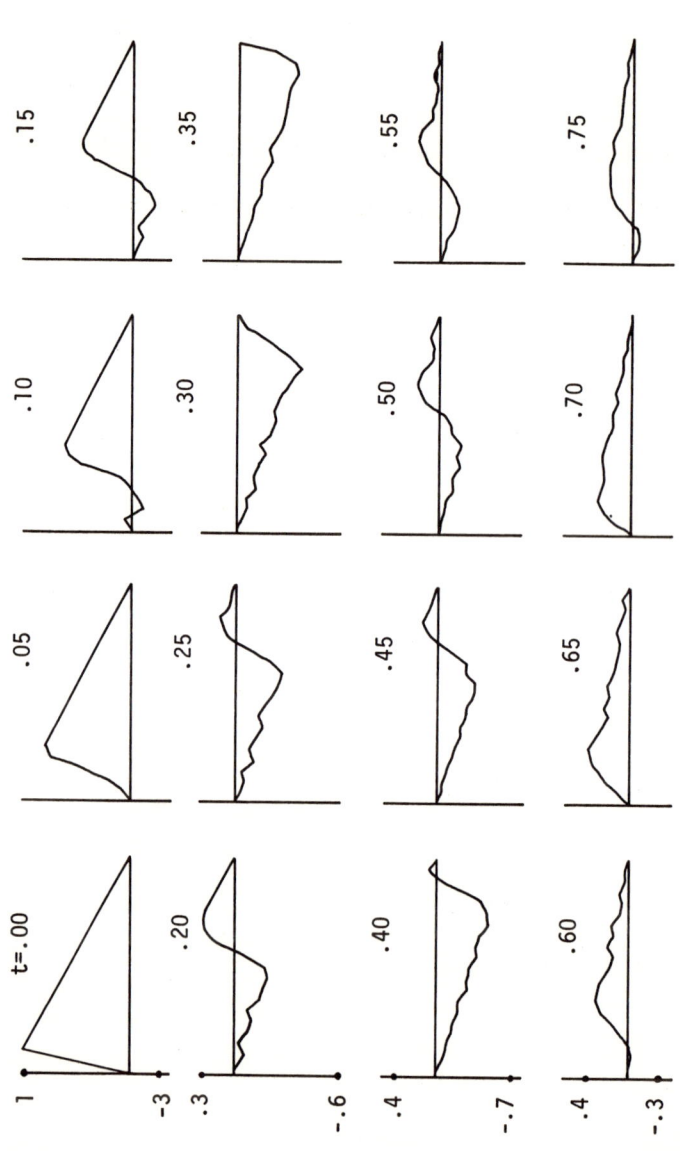

Figure 8.3

n-body problems and applications to such phenomena as the generation of shock waves, see, e.g., Greenspan (12), (14).

Exercises

1. For $t_k = k(0.01)$, $k = 0, 1, \ldots$, and $v_0 = 0$, find the position and velocity at t_{10} of a particle which is in motion on an X axis and whose position x_k at time t_k is given by

   (a) $x_k = (t_k)^2$, $k = 0, 1, 2, \ldots$

   (b) $x_k = 1 + t_k$, $k = 0, 1, 2, \ldots$

   (c) $x_k = \sin(t_k \pi)$, $k = 0, 1, 2, \ldots$

   (d) $x_k = [(-1)^k t_k]/[1 + k]$, $k = 0, 1, 2, \ldots$

   (e) $x_0 = 1$, $x_1 = 0$, $x_2 = 3$, $x_3 = 6$, $x_4 = 1$, $x_5 = -7$, $x_6 = -3$,
   $x_7 = -2$, $x_8 = -6$, $x_9 = -10$, $x_{10} = -15$, $x_{11} = -10$,
   $x_{12} = -4$, $x_{13} = 0$, $x_{14} = 5$, $x_{15} = 7$.

2. A particle's motion on an X axis is given by
   $$a_k + \alpha v_k + \beta f(x_k) = 0, \quad \alpha > 0, \quad k = 0, 1, 2, \ldots,$$
   For $\alpha = 0.3$, $\beta = 1$, $\Delta t = 0.01$, $x_0 = \frac{\pi}{4}$, $f(x_k) = \sin x_k$, generate the motion of the particle up to t_{250} and describe the resulting behavior if $v_0 = 0$.

3. Consider a twenty one particle string with $x_i = \frac{i}{10}$, $i = 0, 1, 2, \ldots, 20$, $\alpha = 0.15$, $\bar{m} = 0.05$, $T_0 = 12.5$, $\Delta t = 0.00025$,

EXERCISES

$\Delta x = 0.1$, $m = 19$, $g = 32.2$, $\varepsilon = 0.01$, and T_1, T_2 defined by (8.25)-(8.26). Place the string in three different initial positions and describe the resulting vibrations.

4. Formulate a discrete model of a liquid which flows out of a canal lock as the lock door is opened and generate the resulting flow on a computer.

5. Formulate a discrete model of heat transfer.

APPENDIX A

MATHEMATICS, THE EXACT SCIENCE

No person can come to the study of mathematics without finding the experience unique. For there are certain qualities about mathematics which make it distinctly different from all other academic disciplines and applied sciences, and it is with these qualities that we shall be concerned.

Let us begin simply, and seemingly quite apart from our subject, by considering some of the extant problems associated with communication between people by means of language. If any particular word, like ship, were flashed on a screen before a large audience, it is doubtful that any two people would form exactly the same mental image of a ship. It follows, similarly, that the meaning of every word is so intimately related to a person's individual experiences, that probably no word has exactly the same meaning to any two people.

To further complicate matters, it does not appear to be possible for anyone to ever find out what a particular word means to anyone else. Suppose for example, that man X asks man Y what the word ship means to him and that man Y replies that a ship is a vessel which moves in, on, or under water. Man X, realizing that even a rowboat tied to a pier is moving by virtue of the earth's rotation, asks man Y to clarify his definition of _ship_ by further defining _to move_. Man Y replies that _to move_ is to relocate from one position to another by such processes as walking, running, driving, flying, sailing, and the like. Man X, for exactness, then asks man Y if by _sailing_ he means the process of navigating a ship which has sails, to which man Y replies yes. "Then", replies man X, "I shall never be able to understand you. You have defined _ship_ in terms of _move_, _move_ in terms of _sailing_, and _sailing_ in terms of _ship_, which was the word originally requiring clarification. You have simply talked around a circle."

The circular process in which men X and Y became involved so quickly is indeed one in which we can all become entangled if we constantly require definitions of words used in definitions. For the total number of words in all existing languages is finite and it would be merely a matter of time to complete a cycle of this verbal merry-go-round.

Now, in constructing the language of mathematical science, the mathematician examines the two semantic problems described above and agrees that no two people will ever completely understand what any particular word means to the other. With this supposition, however, the problem of definitions resulting in a circular process can be, and is, avoided as follows. Suppose, says the mathematician, the words

a	in	path
by	is	point
direction	move	the
fixed	out	trace

are called basic terms and are stated <u>without</u> definition. We all have ideas and feelings about these words, but rather than attempt to make their meanings precise to each other, we shall simply leave them undefined. Now, let us define a <u>line</u> as the path traced out by a point moving in a fixed direction. Note that the word line is defined only in terms of the basic terms. Next, define a <u>plane</u> as the path traced out by a line moving in a fixed direction. Note that <u>plane</u> is defined in terms only of <u>line</u> and of basic terms. Now suppose that man X asks mathematician Y what a <u>plane</u> is. Y responds that a <u>plane</u> is the path traced out by a line moving in a fixed direction. Man X, for clarity, asks mathematician Y what he means precisely by a <u>line</u>, to which Y responds that a <u>line</u> is the path traced out by a point moving in a fixed direction. Man X, seeking futher clarity, asks for the definition of <u>point</u>, to which the mathematician responds, "<u>Point</u> is an undefined basic term", and there the questioning stops.

Thus, every mathematical science begins with basic terms which are undefined and <u>all</u> other concepts are defined by means only of these. <u>Point</u> is an undefined concept of geometry and <u>positive integer</u> is an undefined concept of algebra. No other subject treats its notions this way.

But let us look a bit further into the nature of mathematical concepts. Consider, for example, the geometric concept called a <u>straight line</u>. With a pencil and ruler, we have all at one time or another drawn a straight line. But, indeed, have we really ever drawn a straight line? A mathematical line has <u>no</u> width, while the line we draw with pencil and ruler certainly does have some width, even though one might need a special instrument, like a micrometer, to measure the width. As a matter of fact, the width may even vary as the pencil lead is being used up in the drawing process. Indeed, <u>every</u> physical object has some width and it must follow that the mathematical straight line is an idealized form which exists only in the mind, that is, it is an abstraction. In a similar fashion, it can be shown that <u>all</u> mathematical concepts are idealized forms which exist only in the mind, that is, are abstractions.

So, all mathematical concepts are abstractions which either are undefined or have definitions constructed on basic undefined terms.

After having constructed a system of concepts, the mathematician next seeks a body of rules by which to combine and manipulate his concepts. Thus, mathematical sciences now take on the aspects of a game in that rules of play, which must be followed, have to be enumerated. Each mathematical science has its own rules of play, or, what are technically called assumptions or axioms. The axioms of algebra are indeed quite simple. For example, for the numbers 2, 3 and 5, it is assumed that

$$2 + 5 = 5 + 2$$
$$2 \cdot 5 = 5 \cdot 2$$
$$(2 + 3) + 5 = 2 + (3 + 5)$$
$$(2 \cdot 3) \cdot 5 = 2 \cdot (3 \cdot 5)$$
$$2(3 + 5) = 2 \cdot 3 + 2 \cdot 5 \ .$$

In complete abstract form, then if a,b and c are three positive integers, the algebraist assumes that

$a + b = b + a$	(Commutative axiom of addition)
$a \cdot b = b \cdot a$	(Commutative axiom of multiplication)
$(a + b) + c = a + (b + c)$	(Associative axiom of addition)
$(a \cdot b) \cdot c = a \cdot (b \cdot c)$	(Associative axiom of multiplication)
$a \cdot (b + c) = a \cdot b + a \cdot c$	(Distributive axiom) .

The question which immediately presents itself is how does one go about selecting axioms? Historically, axioms were supposed to coincide with fundamental physical concepts of truth. But, as the nineteenth century chemists and physicists began to destroy the previous year's physical truths, the choice of mathematical axioms became a relative free one. And indeed it is a rather simple matter to show that the axioms stated above for numbers can be false when applied to physical quantities. For example, if a represents sulphuric acid and b represents water, while a + b represents adding sulphuric acid to water and b + a represents adding water to sulphuric acid, then a + b is not equal to b + a, because b + a results in an explosion whereas a + b does not.

In this connection the history associated with the fifth postulate of Euclid is of utmost scientific significance. Indeed, in Euclid's Elements, set forth in about 300 B.C., plane geometry was founded on ten axioms, five of which were called common notions and five of which were called postulates. The axiom of interest to us is the fifth postulate, stated usually in the following equivalent form:

Postulate   Through a point not on a given line one and only one parallel can be drawn to the given line.

Throughout the centuries, Postulate 5 was of serious concern to mathematicians. Euclid himself seems to have avoided its use whenever possible. The reason for its somewhat tenuous position among the other axioms of geometry lay in the realization that it was an assumption about an <u>infinite</u> object, that is, the entire straight line, when science knew of no physical object with any infinite quality or dimension. Indeed, science maintains that everything in the _physical_ world is of finite character. But such a rock of Gibralter was geometry in the realm of mathematics, physics, and astronomy, that instead of seeking a physically acceptable replacement for Postulate 5, mathematicians until the nineteenth century sought primarily to establish its truth.

It was not until the latter part of the eighteenth century and during the nineteenth century that such men as Gauss, Bolyai, and Lobachewsky developed a second geometry in which the postulate was replaced by the assumption that through a point not on a given line one could draw _two_ parallels to the given line. Still later, Riemann developed a third geometry by assuming that _no_ parallels to a given line could be drawn through a point not on the line. And perhaps the greatest scientific impact of these new geometries was that the geometry of Riemann laid the groundwork for the geometry utilized in the Einstein theory of relativity. Thus, mathematical history shows that until some freedom of choice with regard to the selection of axioms was realized, the development of the theory of relativity simply was not possible.

Note, however, that _complete_ freedom of choice in the selection of axioms is available to no man. It would be impractical, for example, to start with assumptions like:

Axiom 1.   All numbers are positive

Axiom 2.   Some numbers are negative,

for these assumptions contradict each other. Indeed, even though a set of axioms does not contain a contradictory pair, it can happen that reasoning from them would yield contradictory conclusions. But, further examination of the very deep problems associated with selection of axioms would be beyond the present scope.

The final difference between mathematics and all other disciplines lies in the reasoning processes allowed in reaching conclusions. There are basically two acceptable types of reasoning in scientific work, inductive reasoning and deductive reasoning. Let us consider each in turn.

Suppose scientist X injects 100 monkeys with virus Y and does not so inject a control group of 100 monkeys. One week later, ninety monkeys in the first group and only five in the second group contract chicken pox. Scientist X, sensing a discovery, repeats the experiment and finds approximately the same statistical results. Further experiments are made in which various environmental factors like heat, light, proximity of cages, and so forth are varied, and in every case X finds that from 85% to 95% of the monkeys receiving virus Y become ill, while only from 3% to 10% of the control group acquire the disease. Scientist X concludes that virus Y is the cause of monkey chicken pox, and the process of reaching his conclusion by experimentation with control is called inductive reasoning. Note that if, after proving his result, X were to inject only one monkey with virus Y, all that he could say would be that the probability is very high that the monkey will become ill. Indeed it is not absolutely necessary that chicken pox will prevail.

Suppose now that mathematician X writes down a set of assumptions, two of which are

    Axiom 1.   All heavenly bodies are hollow.
    Axiom 2.   All moons are heavenly bodies.

Then it __must__ follow, __without__ __exception__, that

    Conclusion: All moons are hollow.

The above type of reasoning from axioms to __necessary__ conclusions is called deductive reasoning. The simple three line argument presented above is called a syllogism. The general process of reaching necessary conclusions from axioms is called deductive reasoning and the syllogism is the fundamental unit in all complex deductive arguments. In mathematics, __all__ conclusions must be reached by deductive reasoning alone. Although, very often __axioms__ are selected after extensive inductive reasoning, no mathematical __conclusion__ can be so reached.

Thus, there is no question of a mathematical conclusion having a high probability of validity as in the case of inductive conclusions. Indeed, if the axioms are absolute truths, then so are the deductive conclusions.

Thus we see that a mathematical science deals with abstract idealized forms which are defined from basic undefined terms, relates its concepts by means of axioms, and establishes conclusions only by deductive reasoning from the axioms. It is the perfect precision of abstract forms and deductive reasoning which makes mathematics an exact science, and it is the prescribing of materials and methods from which one can create meaningful new forms which gives mathematics the form of an art.

APPENDIX B

FORTRAN PROGRAM NAVSTK

Since computer programs for most elementary numerical processes are now readily available in such compendia as that of Carnahan, Luther and Wilkes, we will give in this appendix a typical program for a non-elementary numerical process. The program is called NAVSTK, is given in FORTRAN, and applies to the prototype Navier-Stokes problem of Section 7.2. The program variables are:

OMA = vorticity values
PSI = stream values
N = number of vertical spaces in the grid
M = number of horizontal spaces in the grid
R = Reynold's number
H = grid size
EPS = tolerance for inner-and outer-iterations
C1 = weighting factor for OMA
E1 = weighting factor for PSI
RW = relaxation factor for OMA equations
NM = number of outer-iterations
NCOUNT = number of inner-iterations
W0,W1,W2,W3,W4 = coefficients for the vorticity equation
ISTOP = switch to indicate convergence .

The program itself is as follows:

PROGRAM NAVSTK

```
          PROGRAM NAVSTK
          DIMENSION PSI(50,50),OMA(50,50),SVPSI(50,50),SVOMA(50,50),SVOUT(5
         10,50)
          COMMON N,NPLUS1,M,MPLUS1,NZ,MP
          READ 300,N,M
 300      FORMAT(2I2)
          MPLUS1=M+1
          MMESH=M-1
          JM=0
          NPLUS1=N+1
          NMESH=N-1
          H=1./N
          H2=H*H
          EPS=.001
C     INITIALIZE VECTORS
          NZ=0
          MP=5
          ISTOP=0
          R=50
          RW=1.
          C1=0
          E1=0
 104      CONTINUE
          PRINT 2323,C1
 2323     FORMAT(1H1,F8.2)
          DO 1 I=1,50
          DO 1 J=1,50
          SVOUT(I,J)=0
          SVPSI(I,J)=0
          SVOMA(I,J)=0
          PSI(I,J)=0
 1         OMA(I,J)=0
          NM=0
          E2=1-E1
          C2=1-C1
C     BEGIN LOOP FOR OUTER ITERATIONS
C     SAVE VORTICITY FUNCTION FROM PREVIOUS OUTER ITERATION
 23       DO 40 I=1,NPLUS1
          DO 40 J=1,MPLUS1
 40       SVOUT(I,J)=OMA(I,J)
          NM=NM+1
          NCOUNT=0
C     BEGIN INNER ITERATION FOR STREAM FUNCTION
C       COMPUTE STREAM FUNCTION FOR INNER REGION
 11       DO 2 I=3,NMESH
          DO 2 J=3,MMESH
          SVPSI(I,J)=PSI(I,J)
 2        PSI(I,J)=(-.8*PSI(I,J))+.45*(PSI(I,J-1)+PSI(I,J+1)+PSI(I-1,J)+
         1PSI(I+1,J)+H2*OMA(I,J))
C       COMPUTE STREAM FUNCTION ON TOP AND BOTTOM INNER BOUNDARY LINES
```

```
              DO 3 I=2,N
              PSI(I,2)=(.25*PSI(I,3))
3             PSI(I,M)=.25*PSI(I,MMESH)+.5*H
C       COMPUTE STREAM FUNCTION ON LEFT AND RIGHT INNER BOUNDARY LINES
              DO 4 I=3,MMESH
              PSI(2,I)=   (.25*PSI(3,I))
4             PSI(N,I)=   (.25*PSI(N-1,I))
C       TEST STREAM FUNCTION FOR CONVERGENCE
              DO 5 I=3,NMESH
              DO 5 J=3,MMESH
              DIFF=ABSF(SVPSI(I,J)-PSI(I,J))
              IF(DIFF .GT. EPS) GO TO 6
5             CONTINUE
C       RECALCULATE STREAM FUNCTION USING WEIGHTING
              DO 222 I=3,NMESH
              DO 222 J=3,MMESH
222           PSI(I,J)=E1*SVPSI(I,J)+E2*PSI(I,J)
              DO 114 I=2,NMESH
              IF(PSI(I,M))28,114,114
114             CONTINUE
              GO TO 200
6             NCOUNT=NCOUNT+1
              IF(NCOUNT .GT. 100) GO TO 8
              GO TO 11
C       TEST STREAM FUNCTION FOR DIVERGENCE
8             IF(DIFF .GT. 10) GO TO 28
              PRINT 93
93            FORMAT(1H1,11H PSI VALUES)
              CALL PRNTLST(PSI)
10            FORMAT(10F11.6)
              NCOUNT=0
              GO TO 11
28              PRINT 81
81            FORMAT(13H PSI DIVERGED)
              CALL PRNTLST(PSI)
              CALL PRNTLST(OMA)
              GO TO 699
C       BEGIN INNER ITERATION FOR VORTICITY
200           NCOUNT=0
30              HCONST=C2*(-2./H2)
C       COMPUTE VORTICITY ON BOUNDARY LINES USING WEIGHTING
C       TOP AND BOTTOM BOUNDARY LINES
              DO 12 I=1,NPLUS1
              OMA(I,1)=C1*OMA(I,1)+HCONST*PSI(I,2)
12            OMA(I,M+1)=C1*OMA(I,M+1)+HCONST*(PSI(I,M)-H)
C           LEFT AND RIGHT BOUNDARY LINES
              DO 13 I=2,M
              OMA(1,I)=HCONST*PSI(2,I)+C1*OMA(1,I)
13              OMA(N+1,I)=HCONST*PSI(N,1)+C1*OMA(N+1,I)
90            CONTINUE
```

PROGRAM NAVSTK

```
C         COMPUTE COEFFICIENTS FOR VORTICITY EQUATIONS
C         COMPLETE ONE SWEEP OF INTERIOR
          DO 14 I=2,N
          DO 14 J=2,M
          A1=PSI(I+1,J)-PSI(I-1,J)
          B1=PSI(I,J+1)-PSI(I,J-1)
          A=ABSF(A1)
          B=ABSF(B1)
          W0=4+(A+B)*(R/2)
           IF(A1.GE. 0)15,16
15        W2=1+(R/2)*A
          W4=1
          GO TO 17
16        W2=1
          W4=1+A*(R/2)
17        IF(B1.GE. 0)18,19
18        W1=1
          W3=1+B*(R/2)
          GO TO 20
19        W1=1+B*(R/2)
          W3=1
20        SVOMA(I,J)=OMA(I,J)
          IF(ISTOP .EQ. 1)GO TO 305
          OMA(I,J)=((W1/W0)*OMA(I+1,J)+(W2/W0)*OMA(I,J+1)+(W3/W0)*OMA(I-1,J)
         1+(W4/W0)*OMA(I,J-1))*RW+(1-RW)*OMA(I,J)
          GO TO 14
C         CHECK TO SEE IF DIFFERENCE EQUATIONS ARE SATISFIED TO .001
305       DIFF   =((W1/W0)*OMA(I+1,J)+(W2/W0)*OMA(I,J+1)+(W3/W0)*OMA(I-1,J)
         1+(W4/W0)*OMA(I,J-1))-OMA(I,J)
          DIF=ABSF(DIFF)
          IF(DIF .GT. EPS1)282,14
282       PRINT 183,I,J
          GO TO 700
14        CONTINUE
          IF (ISTOP .EQ. 1) GO TO 700
C     TEST   VORTICITY       FOR CONVERGENCE
          DO 21 I=2,N
          DO 21 J=2,M
          DIFF=ABSF(SVOMA(I,J)-OMA(I,J))
          IF(DIFF .GE. EPS) GO TO 22
21        CONTINUE
C         RECALCULATE VORTICITY USING WEIGHTING
          DO 144 I=2,N
          DO 144 J=2,M
144       OMA(I,J)=C1*SVOMA(I,J)+ C2*OMA(I,J)
          JM=JM+1
C         PRINT OUT EVERY 4 OUTER ITERATES
          IF(JM .EQ. 4)89,59
89        JM=0
          PRINT 79,NM
```

```
      79      FORMAT(1H1,I2,17H OUTER ITERATIONS)
              PRINT 91
              CALL PRNTLST(PSI)
              PRINT 92
              CALL PRNTLST(OMA)
      C       TEST OUTER ITERATIONS FOR CONVERGENCE
      59      CONTINUE
              DO 45 I=1,NPLUS1
              DO 45 J=1,MPLUS1
              DIFF=ABSF(SVOUT(I,J)-OMA(I,J))
              IF(DIFF .GT. EPS) GO TO 7
      45      CONTINUE
              NZ=0
              MP=8
              PRINT 99,NM
      99      FORMAT(1H1,22H PROBLEM CONVERGED IN ,I4)
              PRINT 91
      91      FORMAT(1X,11H PSI VALUES)
              CALL PRNTLST(PSI)
              PRINT 92
      92      FORMAT(1H1,14H OMEGA  VALUES)
              CALL PRNTLST(OMA)
              EPS1=.001
              RMAX=0
              ISTOP=1
      C       CHECK TO SEE IF DIFFERENCE EQUATIONS FOR STREAM FUNCTION ARE SATISFIED
      C       TO A TOLERANCE OF .001
              DO 181 II=3,NMESH
              DO 181 JJ=3,MMESH
              RES=ABSF(PSI(II,JJ)-SVPSI(II,JJ))
              IF(RES .GT. RMAX)301,302
      301     RMAX=RES
      302     CONTINUE
              A=-4*PSI(II,JJ)+PSI(II+1,JJ)+PSI(II,JJ+1)+PSI(II-1,JJ)+PSI(II,JJ-
             11)
              B=-H*H*OMA(II,JJ)
              D=A-B
              IF(D .GT. EPS1) GO TO 182
      181     CONTINUE
              GO TO 90
      182     PRINT 183,II,JJ
      183     FORMAT(1H1,41H DIFFERENCE EQU. NOT SATISFIED AT POINT (,I2,1H,,I2
             1,1H))
              GO TO 699
      C       TEST OUTER ITERATIONS FOR DIVERGENCE
      7       IF(DIFF .GT. 100)199,23
      22      NCOUNT=NCOUNT+1
              IF(NCOUNT .GT. 300) GO TO 24
              GO TO 90
      C       TEST    VORTICITY      FOR DIVERGENCE
```

PROGRAM NAVSTK

```
24      IF(DIFF .GT. 10) GO TO 29
        PRINT 94
94      FORMAT(1H1,13H OMEGA VALUES)
        CALL PRNTLST(OMA)
        PRINT 91
        CALL PRNTLST(PSI)
32      FORMAT(10F11.6)
        NCOUNT=0
        GO TO 90
29      PRINT 82
82      FORMAT(13H OMA DIVERGED)
        CALL PRNTLST(PSI)
        CALL PRNTLST(OMA)
        GO TO 699
199     PRINT 189
189     FORMAT(26H OUTER ITERATIONS DIVERGED)
700     CONTINUE
        PRINT 303,RMAX
303     FORMAT(1H1,17H PSI CONVERGED TO,E12.4)
699     CONTINUE
        SUBROUTINE PRNTLST(Z)
         DIMENSION Z(50,50)
        COMMON N,NPLUS1,M,MPLUS1,NZ,MP
        IF(NZ .EQ. 1) GO TO 103
        IF(N .GT. 11)103,75
75      DO 61 J=1,MPLUS1
        L=MPLUS1-J+1
61      PRINT 52,(Z(I,L),I=1,NPLUS1)
        RETURN
103     NE=0
        NN=11
        DO 51 IP=1,NPLUS1,11
2       NB=IP
        NE=NE+NN
        IF(NE .GT. NPLUS1)101,102
101     NE=NPLUS1
102     DO 51 J=1,MPLUS1
        L=MPLUS1-J+1
        IF(NZ .EQ. 1) GO TO 64
        PRINT 52,(Z(I,L),I=NB,NE)
        GO TO 51
64       PRINT 62,(Z(I,L),I=NB,NE)
62      FORMAT(1X,11E10.2)
51      CONTINUE
52      FORMAT(11F10.5)
        RETURN
        END
```

For additional programs of relative complexity, the reader should consult the Technical Report Series of the Computer Sciences Department, University of Wisconsin.

REFERENCES AND SOURCES FOR FURTHER READING

M. Abramowitz and I. A. Stegun, <u>Handbook of Mathematical Functions</u>, National Bureau of Standards, Washington, D. C., 1965.

B. J. Alder, "Studies in molecular dynamics. III. A mixture of hard spheres," Jour. Chem. Phys., 40, 1964, pp. 2724-2730.

W. F. Ames, <u>Nonlinear Partial Differential Equations in Engineering</u>, Academic Press, New York, 1965.

A. A. Amsden, "The particle-in-cell method for the calculation of the dynamics of compressible fluids," Rpt. 3406, Los Alamos Scientific Lab., L. A., N. M., 1966.

P. M. Anselone (Ed.), <u>Nonlinear Integral Equations</u>, University of Wisconsin Press, Madison, Wis., 1964.

G. K. Batchelor, "On steady laminar flow with closed streamlines at large Reynolds numbers," Jour. Fluid Mech., 1, 1956, pp. 177-190.

H. Bateman, <u>Partial Differential Equations of Mathematical Physics</u>, Cambridge Univ. Press, Cambridge, 1964.

R. E. Bellman, R. E. Kalaba, and J. A. Lockett, <u>Numerical Inversion of the Laplace Transform</u>, Elsevier, N. Y., 1966.

I. S. Berezin and N. P. Zhidkov, <u>Computing Methods</u>, vols. I and II, Addison-Wesley, Reading, Mass., 1965.

P. W. Berg and J. L. McGregor, <u>Elementary Partial Differential Equations</u>, Holden-Day, San Francisco, 1966.

R. L. Berger and N. Davids, "General computer method analysis of condition and diffusion in biological systems with distributive sources," Rev. Sci. Instr., 36, 1965, pp. 88-93.

D. Bernstein, <u>Existence Theorems in Partial Differential Equations</u>, Princeton Univ. Press, Princeton, N. J., 1950.

L. Bers, "On mildly nonlinear partial differential equations of elliptic type," Jour. Res. N. B. S., 51, 1953, pp. 229-236.

REFERENCES

L. Bieberbach, *Theorie der Differentialgleichungen*, Dover, New York, 1944.

A. D. Booth, *Numerical Methods*, Butterworths, London, 1955.

R. T. Boughner, "The discretization error in the finite difference solution of the linearized Navier-Stokes equations for incompressible fluid flow at large Reynolds number," TM 2165, Oak Ridge Nat. Lab., Oak Ridge, Tenn., 1968.

J. H. Bramble, "Error estimates for difference methods in forced vibration problems," SIAM Jour. Num. Anal., 3, 1966, pp. 1-12.

A. Carasso and S. V. Parter, "An analysis of 'Boundary-Value Techniques' for parabolic problems," Math. Comp., 110, 1970, pp. 315-340.

B. Carnahan, H. A. Luther, and J. O. Wilkes, *Applied Numerical Methods*, Wiley, N. Y., 1969.

J. W. Carr III, "Error bounds for the Runge-Kutta single step integration process," Jour. ACM, 5, 1958, pp. 39-44.

L. Cesari, *Asymptotic Behavior and Stability Problems in Ordinary Differential Equations*, Springer, Berlin, 1959.

A. J. Chorin, "The numerical solution of the Navier-Stokes equations for an incompressible fluid," Bull. AMS, 73, 1967, pp. 928-931.

R. V. Churchill, *Fourier Series and Boundary Value Problems*, McGraw-Hill, New York, 1941.

C. W. Clenshaw, "The solution of van der Pol's equation in Chebychev series," in *Numerical Solutions of Nonlinear Differential Equations*, Wiley, N. Y., 1966, pp. 55-63.

E. A. Coddington and N. Levinson, *Theory of Ordinary Differential Equations*, McGraw-Hill, N. Y., 1955.

L. Collatz, (1) *The Numerical Treatment of Differential Equations*, Springer-Verlag, Berlin, 1959. (2) *Functional Analysis and Numerical Mathematics*, Academic Press, N. Y., 1966.

P. Concus, "Numerical solution of the minimal surface equation," Math. Comp., 21, 1967, pp. 340-350.

R. Courant and K. O. Friedrichs, Supersonic Flow and Shock Waves, Interscience, N. Y., 1948.

R. Courant, K. Friedrichs and H. Lewy, "Uber die partiellen Differenzengleichungen der Mathematischen Physik," Math. Ann., 1928, pp. 32-74.

R. Courant and D. Hilbert, Methods of Mathematical Physics, Vol. II, Interscience, N. Y., 1962.

R. Courant, E. Isaacson and M. Rees, "On the solution of nonlinear hyperbolic differential equations by finite differences," Comm. Pure Appl. Math., 5, 1952, pp. 243-255.

C. W. Cryer, "Stability analysis in discrete mechanics," Tech. Rpt. #67, Dept. of Computer Sciences, Univ. of Wis., Madison, 1969.

J. M. A. Danby, Fundamentals of Celestial Mechanics, Macmillan, N. Y., 1962.

J. W. Daniel, The Approximate Minimization of Functionals, Prentice Hall, Englewood Cliffs, N. J., 1971.

D. F. Davidenko, "Construction of difference equations for approximating the solution of the Euler-Poisson-Darboux equation," (in Russian), Dokl. Akad. Nauk SSSR, 142, 1962, pp. 510-513.

H. T. Davis, Introduction to Nonlinear Differential and Integral Equations, U. S. Atomic Energy Commission, Washington, D. C., 1960.

P. J. Davis and P. Rabinowitz, Numerical Integration, Blaisdell, Waltham, Mass., 1967.

C. R. Deeter and G. Springer, "Discrete harmonic kernels," Jour. Math. and Mech., 14, 1965, pp. 413-438.

J. Douglas, Jr., "A survey of numerical methods for parabolic differential equations," in Advances in Computers, II, Academic Press, N. Y., 1961, pp. 1-55.

REFERENCES

S. E. Dreyfus, "The numerical solution of nonlinear control problems," in *Numerical Solutions of Nonlinear Differential Equations*, Wiley, N. Y., 1966, pp. 335-363.

R. J. Duffin, "Basic properties of discrete analytic functions," Duke Math. Jour., 23, 1956, pp. 335-363

M. Esser, *Differential Equations*, Saunders, Philadelphia, 1968.

D. K. Faddeev and V. N. Faddeeva, *Computational Methods of Linear Algebra*, Freeman, San Francisco, 1963.

E. Fehlberg, "Classical fifth-, sixth-, seventh-, and eighth-order Runge-Kutta formulas with stepsize control," TR T-287, NASA, 1968.

E. Fermi, J. R. Pasta, and S. Ulam, "Studies of nonlinear problems. I," Rpt. #1940, Los Alamos Scientific Lab., L. A., N. M., 1955.

G. E. Forsythe, "Solving linear algebraic equations can be interesting," Bull. AMS, 59, 1953, pp. 299-329.

G. E. Forsythe and W. R. Wasow, *Finite-Difference Methods for Partial Differential Equations*, Wiley, New York, 1960.

L. Fox, (1) *The Numerical Solution of Two-Point Boundary Problems in Ordinary Differential Equations*, Oxford Univ. Press, Fairlawn, N. J., 1957. (2) (Editor) *Numerical Solution of Ordinary and Partial Differential Equations*, Addison-Wesley, Reading, Mass., 1962. (3) *An Introduction to Numerical Linear Algebra*, Oxford Univ. Press, N. Y., 1965.

J. N. Franklin, "Difference methods for stochastic ordinary differential equations," Math. Comp., 19, 1965, pp. 552-561.

A. Friedman, *Partial Differential Equations of Parabolic Type*, Prentice Hall, Englewood Cliffs, N. J., 1964.

C. E. Froberg, *Introduction to Numerical Analysis*, Addison-Wesley, Reading, Mass., 1965.

REFERENCES

J. E. Fromm, "Lectures on large scale finite difference computation of incompressible fluid flows," Rpt. RJ 617, IBM Research, San Jose, Calif., 1969.

J. E. Fromm and F. H. Harlow, "Numerical solution of the problem of vortex street development," Phys. of Fluids, 6, 1963, pp. 975-985.

R. A. Gangolli and D. Ylvisaker, Discrete Probability, Harcourt, Brace and World, New York, 1967.

P. R. Garabedian, Partial Differential Equations, Wiley, New York, 1964.

H. Geiringer, "On the solution of systems of linear equations by certain iteration methods," Reissner Anniv. Vol., Ann Arbor, Mich., 1949, pp. 365-393.

S. K. Godunov and V. S. Ryabenki, The Theory of Difference Schemes, Wiley, N. Y., 1964.

H. Goldstein, Classical Mechanics, Addison-Wesley, Reading, Mass., 1959.

M. Golomb and H. F. Weinberger, "Optimal approximation and error bounds," in On Numerical Approximation, Univ. Wisconsin Press, Madison, Wis., 1959, pp. 117-191.

E. T. Goodwin (Ed.), Modern Computing Methods, Philosophical Library, 1961.

D. Greenspan, (1) Theory and Solution of Ordinary Differential Equations, Macmillan, N. Y., 1960. (2) Introduction to Partial Differential Equations, McGraw-Hill, N. Y., 1961. (3) Introductory Numerical Analysis of Elliptic Boundary Value Problems, Harper and Row, N. Y., 1965. (4) (Editor), Numerical Solutions of Nonlinear Differential Equations, Wiley, N. Y., 1966. (5) Introduction to Calculus, Harper and Row, N. Y., 1968. (6) Lectures on the Numerical Solution of Linear, Singular and Nonlinear Differential Equations, Prentice-Hall, Englewood Cliffs, N. J., 1969. (7) "Resolution of classical capacity problems by means of a digital computer," Can. Jour. Phys., 44, 1966, pp. 2605-2613. (8) "On

REFERENCES

approximating extremals of functionals. Part II. Theory and generalizations related to boundary value problems for nonlinear differential equations," Int. Jour. Eng. Sci., 5, 1967, pp. 571-588. (9) "Numerical solution of a class of nonsteady cavity flow problems," BIT, 8, 1968, pp. 287-294. (10) "Numerical studies of prototype cavity flow problems," The Comp. Jour., 12, 1969, pp. 89-94. (11) "Computer simulation of transverse string vibrations," BIT, 11, 1971, pp. 399-408. (12) "Numerical studies of the 3-body problem," SIAM Jour. Appl. Math., 20, 1971, pp. 67-78. (13) Introduction to Numerical Analysis and Applications, Markham Publishers, Chicago, 1971. (14) "Discrete liquid flow," Tech. Rpt. #14, Computing Center, Univ. Wis., Madison, 1970.

D. Greenspan and P. C. Jain, "Numerical study of subsonic fluid flow by a combination variational integral-finite difference technique," Jour. Math. Anal. Appl., 18, 1967, pp. 85-111.

D. Greenspan and S. V. Parter, "Mildly nonlinear elliptic partial differential equations and their numerical solution, II," Num. Math., 7, 1965, pp. 129-146.

D. Greenspan and P. Werner, "A numerical method for the exterior Dirichlet problem for the reduced wave equation," Arch. Rat. Mech. Anal., 23, 1966, pp. 288-316.

W. Grobner, Contributions to the Method of Lie Series, B. I. Hochschulskripten 802/802a*, Bibliographisches Institut, Mannheim.

R. W. Hamming, Numerical Methods for Scientists and Engineers, McGraw-Hill, N. Y., 1962.

K. F. Hansen, B. V. Koen and W. W. Little, Jr., "Stable numerical solutions of the reactor kinetics equations," Nuclear Sci. Eng., 22, 1965, pp. 51-59.

D. R. Hartree, Numerical Analysis, Oxford Univ. Press, Oxford, 1952.

W. Heinsenberg, Physics and Philosophy, Harper and Row, N. Y., 1958.

W. J. Hemmerle, Statistical Computations on a Digital Computer, Blaisdell, Waltham, Mass., 1967.

P. Henrici, (1) <u>Discrete Variable Methods in Ordinary Differential Equations</u>, Wiley, N. Y., 1962. (2) <u>Elements of Numerical Analysis</u>, Wiley, N. Y., 1964.

K. Heun, "Neue Methode zur approximativen Integration der Differentialgleichungen einer unabhangigen Variable," ZAMP, 45, 1900, pp. 23-38.

F. B. Hildebrand, (1) <u>Introduction to Numerical Analysis</u>, McGraw-Hill, N. Y., 1953, (2) <u>Finite-Difference Equations and Simulations</u>, Prentice-Hall, Englewood Cliffs, N. J., 1968.

J. O. Hirschfelder, C. F. Curtiss and R. B. Bird, <u>Molecular Theory of Gases and Liquids</u>, Wiley, N. Y., 1954.

H. Hochstadt, <u>Special Functions of Mathematical Physics</u>, Holt, Rinehart and Winston, N. Y., 1961.

R. W. Hockney, "A fast direct solution of Poisson's equation using Fourier analysis," Jour. ACM, 12, 1965, pp. 95-113.

L. Hormander, <u>Linear Partial Differential Operators</u>, Academic Press, N. Y., 1963.

A. S. Householder, (1) <u>Principles of Numerical Analysis</u>, McGraw-Hill, New York, 1953. (2) <u>The Theory of Matrices in Numerical Analysis</u>, Blaisdell, N. Y., 1964.

A. Huber, "Some results in generalized axially symmetric potentials," Proc. Conf. Diff. Eqs., Coll. Pk., Md., 1955, pp. 147-155.

E. Isaacson and H. B. Keller, <u>Analysis of Numerical Methods</u>, Wiley, N. Y., 1966.

N. N. Janenko (Editor), <u>Difference Methods for Solutions of Problems in Mathematical Physics</u>, Amer. Math. Soc., Providence, R. I., 1967.

F. John, (1) "On integration of parabolic equations by difference methods," Comm. Pure Appl. Math., 5, 1952, pp. 155-211. (2) <u>Lectures on Advanced Numerical Analysis</u>, Gordon and Breach, N. Y., 1967.

REFERENCES

E. Kamke, Differentialgleichungen, Vols. I and II, Akad. Verlag., Leipzig, 1959.

M. Kawaguti, "Numerical solution of the Navier-Stokes equations for the flow in a two dimensional cavity," Jour. Phys. Soc., Japan, 16, 1961, pp. 2307-2315.

H. B. Keller, Numerical Methods for Two-Point Boundary Value Problems, Blaisdell, Waltham, Mass., 1968.

O. D. Kellogg, Foundations of Potential Theory, Dover, N. Y., 1953.

L. G. Kelly, Handbook of Numerical Methods and Applications, Addison-Wesley, Reading, Mass., 1967.

J. G. Kemeny and J. L. Snell, Mathematical Models in the Social Sciences, Ginn and Co., N. Y., 1962.

J. Kowalik and M. R. Osborne, Methods for Unconstrained Optimization Problems, Elsevier, N. Y., 1968.

H. O. Kreiss, "Difference approximations for the initial-boundary value problem for hyperbolic differential equations," in Numerical Solutions of Nonlinear Differential Equations, Wiley, N. Y., 1966, pp. 141-166.

K. S. Kunz, Numerical Analysis, McGraw-Hill, N. Y., 1957.

W. Kutta, "Beitrag zur naherungsweisen Integration totaler Differentialgleichungen," ZAMP, 46, 1901, pp. 435-453.

O. A. Ladyzhenskaya, The Mathematical Theory of Viscous Incompressible Flow, 2nd Ed., Gordon and Breach, N. Y., 1969.

J. D. Lawson, "An order six Runge-Kutta process with extended region of stability," SIAM Jour. Num. Anal., 4, 1967, pp. 620-625.

P. D. Lax, "Nonlinear partial differential equations and computing," SIAM Review, 11, 1969, pp. 7-19.

P. D. Lax and B. Wendroff, "Systems of conservation laws," Comm. Pure Appl. Math., 13, 1960, pp. 217-237.

C. E. Leith, "Numerical hydrodynamics of the atmosphere," in Proc. Symp. Applied Math. of Amer. Math. Soc., Amer. Math. Soc., Providence, R. I., 1967, pp. 125-137.

H. Levy and F. Lessman, *Finite Difference Equations*, Macmillan, N. Y., 1961.

H. A. Luther and H. P. Konen, "Some fifth-order classical Runge-Kutta formulas," SIAM Rev., 7, 1965, pp. 551-558.

M. M. May and R. H. White, "Hydrodynamic calculation of general relativistic collapse," The Physical Review, 141, 1966, pp. 1232-1241.

N. W. McLachlan, *Ordinary Non-Linear Differential Equations in Engineering and Physical Sciences*, Oxford Univ. Press, London, 1950.

P. K. Mehta, "Cylindrical and spherical elastoplastic stress waves by a unified direct analysis method," AIAA Jour., 5, 1967, pp. 2242-2248.

W. G. Melbourne, "Lunar and planetary flight mechanics," Jet Propulsion Laboratory, Pasadena, Calif., January, 1968.

R. H. Miller and N. Alton, "Three dimensional n-body calculations," ICR Quart. Rpt. #18, Univ. Chicago, Chicago, 1968.

R. D. Mills, "On a closed motion of a fluid in a square cavity," Jour. Roy. Aer. Soc., 69, 1965, pp. 116-120.

W. E. Milne, (1) *Numerical Calculus*, Princeton Univ. Press, Princeton, N. J., 1950. (2) *Numerical Solution of Differential Equations*, Wiley, New York, 1953.

C. L. Miracle, "Approximate solutions of the telegrapher's equation by difference equation methods," Jour. SIAM, 10, 1962, pp. 517-527.

R. von Mises, *Mathematical Theory of Compressible Fluid Flow*, Academic Press, New York, 1958.

R. E. Moore, *Interval Analysis*, Prentice Hall, Englewood Cliffs, N. J., 1966.

R. H. Moore, "A Runge-Kutta procedure for the Goursat problem in hyperbolic partial differential equations," Arch. Rat. Mech. Anal., 1, 1961, pp. 37-63.

REFERENCES

P. M. Morse and H. Feshbach, <u>Methods of Theoretical Physics</u>, Vols. I and II, McGraw-Hill, N. Y., 1953.

Z. Nehari, <u>Conformal Mapping</u>, McGraw-Hill, N. Y., 1952.

J. von Neumann, "Proposal and analysis of a new numerical method for the treatement of hydrodynamical shock problems," in <u>Collected Works of John von Neumann</u>, VI, Pergamon, N. Y., 1963, pp. 361-379.

B. Noble, <u>Numerical Methods</u>, Oliver and Boyd, London, 1964.

W. F. Noh, "CEL: a time-dependent, two-space-dimensional, coupled Eulerian-Lagrange code," in <u>Methods in Computational Physics</u>, N. Y., 1964, pp. 117-180.

V. V. Novozhilov, <u>Foundations of the Nonlinear Theory of Elasticity</u>, Graylock, Rochester, N. Y., 1953.

A. M. Ostrowski, <u>Solution of Equations and Systems of Equations</u>, 2nd edition, Academic Press, N. Y., 1966.

F. Pan and A. Acrivos, "Steady flows in rectangular cavities," Jour. Fluid Mech., 28, 1967, pp. 643-655.

C. E. Pearson, "A computational method for viscous flow problems," Jour. Fluid Mech., 21, 1965, pp. 611-622.

I. G. Petrovsky, <u>Partial Differential Equations</u>, Interscience, N. Y., 1954.

N. A. Phillips, "Numerical weather prediction," in <u>Advances in Computers</u>, I, Academic Press, N. Y., 1960, pp. 43-91.

S. T. Pohozaev, "The Dirichlet problem for the equation $\Delta u = u^2$," Soviet Math., 1, 1960, pp. 1143-1146.

G. Polya and G. Szego, <u>Isoperimetric Inequalities in Mathematical Physics</u>, Princeton Univ. Press, Princeton, N. J., 1951.

R. W. Preisendorfer, <u>Radiative Transfer on Discrete Spaces</u>, Pergamon, N. Y., 1965.

L. B. Rall, (1) (Editor) Error in Digital Computation, Vol. I and II, Wiley, N. Y., 1965. (2) Computational Solution of Nonlinear Operator Equations, Wiley, N. Y., 1969.

A. Ralston, A First Course in Numerical Analysis, McGraw-Hill, N. Y., 1965.

A. Ralston and H. S. Wilf (Editors), Mathematical Methods for Digital Computers, Wiley, N. Y., 1960.

P. L. Richman, "ε-Calculus," TR 105, Dept. Comp. Sci., Stanford Univ., 1968.

R. D. Richtmyer and K. W. Morton, Difference Methods for Initial-Value Problems, 2nd edition, Interscience, N. Y., 1967.

D. J. Rose, "An algorithm for solving a special class of tridiagonal systems of linear equations," Comm. ACM, 12, 1969, pp. 234-236.

W. C. Royster, "A Poisson integral formula for the ellipse and some applications," Proc. AMS, 15, 1964, pp. 661-670.

M. Sajben, "An exact solution for axially symmetric equilibrium electron density distributions," Phys. Fluids, 11, 1968, pp. 2501-2502.

C. Saltzer, "Discrete potential theory and boundary value problems," Duke Math. Jour., 31, 1964, pp. 299-320.

D. Sarafyan, "Seventh order, 10-stage Runge-Kutta formulas," Tech. Rpt. 38, Math. Dept., LSU, New Orleans, 1970.

M. Schechter, "On the Dirichlet problem for second order elliptic equations with coefficients singular on the boundary," Comm. Pure Appl. Math., 13, 1960, pp. 321-328.

S. Schechter, "Iteration methods for nonlinear problems," Trans. Amer. Math. Soc., 1962, pp. 179-189.

H. Schlichting, Boundary Layer Theory, McGraw-Hill, N. Y., 1960.

D. Schultz, Experimental Numerical Solution of the Navier-Stokes Equations for Flow of a Fluid in a Heated, Closed, Cavity, Ph.D. Thesis, Dept. Comp. Sci., Univ. Wis., Madison, 1970.

REFERENCES

G. H. Shortley, R. Weller, P. Darbey and E. H. Gamble, "Numerical solution of axisymmetrical problems, with applications to electrostatics and torsion," Eng. Exp. Sta. Bull. No. 128, Ohio State Univ., 1947.

J. W. Siry, J. P. Murphy and I. J. Cole, "The Goddard general orbit determination system," NASA-TM-X-63413; X-550-68-218, NASA Goddard Space Flight Ctr., Greenbelt, Md., 1968.

G. D. Smith, Numerical Solution of Partial Differential Equations, Oxford Univ. Press, N. Y., 1965.

J. Smith, "The coupled equation approach to the numerical solution of the biharmonic equation by finite differences," Parts I and II, SIAM Jour. Num. Anal., 5, 1968, pp. 323-339; 7, 1970, pp. 104-111.

M. Soare, Application of Finite Difference Equations to Shell Analysis, Pergamon, N. Y., 1967.

I. S. Sokolnikoff and R. M. Redheffer, Mathematics of Physics and Modern Engineering, McGraw-Hill, N. Y., 1958.

R. W. Southworth and S. L. DeLeeuw, Digital Computation and Numerical Methods, McGraw-Hill, N. Y., 1965.

D. B. Spalding, Convective Mass Transfer, An Introduction, McGraw-Hill, N. Y., 1963.

R. G. Stanton, Numerical Methods for Science and Engineering, Prentice Hall, Englewood Cliffs, N. J., 1961.

J. J. Stoker, Water Waves, Interscience, N. Y., 1957.

D. J. Struik, Lectures on Classical Differential Geometry, Addison-Wesley, Reading, Mass., 1950.

J. L. Synge, (1) The Hypercircle in Mathematical Physics, Cambridge Univ. Press, Cambridge, 1957. (2) Relativity: The Special Theory, North-Holland, Amsterdam, 1965.

E. F. Taylor and J. A. Wheeler, Spacetime Physics, Freeman, San Francisco, 1966.

C. B. Tompkins and W. L. Wilson, Jr., *Elementary Numerical Analysis*, Prentice Hall, Englewood Cliffs, N. J., 1969.

J. Todd (Editor), *Survey of Numerical Analysis*, McGraw-Hill, N. Y., 1962.

F. G. Tricomi, *Integral Equations*, Interscience, N. Y., 1957.

M. Urabe, "Numerical study of periodic solutions of the van der Pol equation," in *International Symposium on Nonlinear Differential Equations and Nonlinear Mechanics*, Academic Press, N. Y., 1963, pp. 184-192.

R. S. Varga, *Matrix Iterative Analysis*, Prentice-Hall, Englewood Cliffs, N. J., 1962.

C. H. Warlick and D. M. Young, "A priori estimates for the determination of the optimum relaxation factor for the successive overrelaxation method," TR 105, Comp. Ctr., U. Texas, Austin, 1970.

B. Wendroff, *Theoretical Numerical Analysis*, Academic Press, N. Y., 1966.

G. J. Whitrow, *The Natural Philosophy of Time*, Harper and Row, N. Y., 1961.

J. H. Wilkinson, *Rounding Errors in Algebraic Processes*, Prentice-Hall, Englewood Cliffs, N. J., 1963.

N. J. Zabusky, "Elastic solution for the vibrations of a nonlinear continuous model string," Jour. Mathematical Phys., 3, 1962, pp. 1028-1039.

M. Zlamal, "On the finite element method," Num. Mat., 12, 1968, pp. 394-409.

ANSWERS TO SELECTED EXERCISES

Chapter 1

3. a, b, d, f
4. a - f
5. b, d
9. (a) $\omega = 1.8$, $x_1^{(4)} = 2.23$, $x_2^{(4)} = 0.69$, $x_3^{(4)} = 2.07$
   $\omega = 1.4$, $x_1^{(4)} = 2.10$, $x_2^{(4)} = 0.39$, $x_3^{(4)} = 1.05$
   $\omega = 1.0$, $x_1^{(4)} = 2.13$, $x_2^{(4)} = 0.39$, $x_3^{(4)} = 1.03$
   $\omega = 0.6$, $x_1^{(4)} = 1.85$, $x_2^{(4)} = 0.28$, $x_3^{(4)} = 0.79$
   $\omega = 0.2$, $x_1^{(4)} = 0.97$, $x_2^{(4)} = 0.06$, $x_3^{(4)} = 0.19$

   (c) $\omega = 1.8$, $x_1^{(4)} = 0.90$, $x_2^{(4)} = 0.76$, $x_3^{(4)} = -0.55$
   $\omega = 1.4$, $x_1^{(4)} = 0.19$, $x_2^{(4)} = 1.14$, $x_3^{(4)} = -0.93$
   $\omega = 1.0$, $x_1^{(4)} = 0.15$, $x_2^{(4)} = 1.17$, $x_3^{(4)} = -0.94$
   $\omega = 0.6$, $x_1^{(4)} = 0.17$, $x_2^{(4)} = 1.02$, $x_3^{(4)} = -0.82$
   $\omega = 0.2$, $x_1^{(4)} = 0.11$, $x_2^{(4)} = 0.52$, $x_3^{(4)} = -0.35$.

Chapter 2

2. (a) exact solution: $y = \sin x$
   (b) exact solution: $y = 2e^x - e^{2x}$
   (c) exact solution: $y = 1 + \frac{15}{32}x - \frac{1}{16}x^2 - \frac{1}{12}x^3$

4. (a) $y_1 = 0.479$, $y_2 = 0.841$, $y_3 = 0.997$

(c) $y_1 = 2.320$, $y_2 = 11.594$, $y_3 = 77.789$

(e) $y_1 = 1.361$, $y_2 = 2.833$, $y_3 = 17.077$

(g) $y_1 = 0.807$, $y_2 = 0.843$, $y_3 = 0.933$

5. (a) $\Delta x > 2$, (b) $\Delta x > .02$, (c) $\Delta x > .002$

7. (a) exact solution: $y = -\sqrt{3} \sin x + \cos x$

(b) exact solution: $y = (e - e^{-4})^{-1} [e^x - e^{-4x}]$

(c) general solution: $y = c_1 e^x + c_2 e^{4x} + \frac{1}{4}x^2 + \frac{1}{8}x + \frac{9}{32}$

(f) general solution: $y = c_1 e^{x^2} + c_2 x e^{x^2}$.

Chapter 3

1. (a) elliptic   (d) hyperbolic
    (b) hyperbolic   (e) hyperbolic
    (c) parabolic   (f) hyperbolic.

2. (a) hyperbolic on the upper-half plane, elliptic on the lower-half plane, parabolic on the X-axis.

(b) hyperbolic outside the unit circle $x^2 + y^2 = 1$, elliptic inside, and parabolic on the circle.

5. $u(2,2) \sim -2$, $u(4,2) \sim 0$, $u(2,4) \sim -6$.

9. (a) $S^i: \xi^2 + \eta^2 = 1$, $F(\xi,\eta) = 1$

SELECTED ANSWERS

14. $C \sim 4.26$.

Chapter 4

1. (a) $u(\frac{1}{4}, 1) \sim -16800$, $u(\frac{1}{2}, 1) \sim 23759$, $u(\frac{3}{4}, 1) \sim -16800$.

Chapter 5

2. (a) $u = 1 - t$
   (b) $u = x + \frac{1}{6}[(x+t)^3 - (x-t)^3]$
   (c) $u = \frac{1}{2}[(x+t)^2 + (x-t)^2 + \int_{x-t}^{x+t} e^{-r^2} dr]$

3. $-3 \leq x \leq 3$, $-2 \leq x \leq 4$, $-6 \leq x \leq 0$, $-1 \leq x \leq 15$, $-8 \leq x \leq -6$, $-9 \leq x \leq 3$.

4. $y - 3 = \pm x$, $y - 3 = \pm(x-1)$, $y - 3 = \pm(x+3)$, $y - 8 = \pm(x-7)$, $y + 1 = \pm(x+7)$, $y - 6 = \pm(x+3)$

Chapter 6

1. (a) $2, \frac{7}{3}, \frac{14}{5}$
   (c) $\frac{4}{3}, \frac{19}{12}, 2$

2. (a) $-2y'' = 0$
   (c) $x - 2y'' = 0$
   (d) $2y'' + y^3(5+4x) = 0$

3. $y(\frac{1}{4}) \sim 0.25$, $y(\frac{1}{2}) \sim 0.5$, $y(\frac{3}{4}) \sim 0.75$.

308   SELECTED ANSWERS

5. $y(\frac{1}{4}) \sim 0.40$, $y(\frac{1}{2}) \sim 0.80$, $y(\frac{3}{4}) \sim 1.06$.

10. (a) $u_{xx} + u_{yy} = \frac{1}{2}e^u$

    (b) $u_{xx} + u_{yy} = u$

    (c) $u_{xx} + u_{yy} = H(u)$

    (d) $(1 + u_y^2)u_{xx} - 2u_x u_y u_{xy} + (1 + u_x^2)u_{yy} = 0$

Chapter 7

3. Two primary vortices for relatively small $\mathcal{R}$.

7. All are hyperbolic.

Chapter 8

1. (a) $x_{10} = 0.01$, $v_k = -v_{k-1} + \frac{2}{(0.01)^3}(2k-1)$

   (b) $x_{10} = 1.1$

   (c) $x_{10} = \sin\frac{\pi}{10}$

   (d) $x_{10} = \frac{1}{110}$

   (e) $x_{10} = -15$, $v_{10} = 400$.

SUBJECT INDEX

Abstraction 279
Acceleration 261-262
Algebraic system 1-22, 64, 108, 150, 170, 194, 231
Arithmetic mean 83
Axiom 280

Backward difference 28, 61
Biharmonic equation 227
Boundary condition 26, 59, 119, 145, 154, 164, 186, 193, 209
Boundary function 75, 78, 102
Boundary grid point 86, 99, 121, 230
Boundary lattice point 86, 99, 121, 230
Boundary value problem 26, 57-65, 66, 73-80, 141, 143, 144, 180, 185, 190, 195-196
Boundary value technique 140-150, 175-182, 229-236
Burger's equation 149

CDC 3600 194
Capacity 102-106
Cauchy problem 153-160
Cavity flow 208, 228
Central difference 28, 143, 172, 196
Centroid 260
Characteristics 159, 164, 182, 242, 244, 247
Compressible 208
Conformal map 78
Conservation of energy 264-267
Conservative form 255
Coupled system 212
Cramer's rule 2
Crank-Nicolson method 134-138

Damped motion 44, 262-264
Damping 44, 262-264
Deductive reasoning 283
Density 246

Detatched wave 182
Determinant 242
Diagonal dominance 3, 10, 60, 61, 90, 98, 131, 148, 149, 218-219, 223
Difference 28, 61
Difference analogue 141, 175, 179
Difference approximation 149, 160, 161, 177, 181, 212, 250
Difference equation 49, 58, 59, 63, 80-83, 97, 107, 138, 231, 232
Differential equation
  biharmonic 227
  Burger's 149
  elliptic 70-113, 212
  Euler 189, 190, 197, 199
  gas dynamical 71-72
  heat 71, 118-138
  hyperbolic 70, 71, 153-182
  Laplace 70, 73, 80, 98, 112
  linear partial 69, 95-98, 113
  mildly nonlinear 70, 106-113, 138-150, 174-182
  minimal surface 71
  Navier-Stokes 208-209, 239-240
  Newton's dynamical 43
  ordinary 26-66, 143, 180, 189
  parabolic 70, 71, 118-150, 237, 239
  partial 69, 212
  potential 70
  quasilinear 69, 241, 246, 249, 252
  radiation 64
  reduced wave 182
  soap film 7, 199
  van der Pol's 52-56
  wave 71, 153-182, 243

Dirichlet problem 75-79, 83, 86, 98, 107, 109, 111, 198
Discrete model 259-273
Double precision 56
Double sequence 212
Dynamical equation 259, 262

Electrostatic capacity 102-106
Elliptic partial differential equation 70-113, 212
Euler equation 189, 190, 197, 199
Euler's method 33-34, 47
Explicit method 129, 165, 237-238
Exterior Dirichlet problem 79-80, 91-94, 99, 103
Extremization of functionals 185-204

Force 44
Formula of D'Alembert 154, 155
Forward difference 28, 61, 191, 196, 262
Forward-backward technique 62, 98, 219
Fourier integral 120
Fourier series 76, 78, 120, 154
Functional 186, 190, 193

Gas 71-72, 208, 240
Gas dynamical equation 71-72
Gauss elimination 5-10, 11
Generalized Newton's formula 16
Generalized Newton's method 12-22, 65, 139, 145, 149, 150, 181, 194, 218, 231
Generalized solution 210-211
Geodesic 192-195
Gravity 44, 264, 270
Grid point 28, 29, 34, 85, 121
Grid size 85

Half-plane 119, 154
Harmonic function 73, 96, 107, 126
Heat equation 71, 118-138
Heun's formulas 38

Hyperbolic partial differential equation 70, 71, 153-182
Hyperbolic system of partial differential equations 240-249

Implicit method 130, 136, 168, 171
Incompressible 208
Inductive reasoning 283
Initial condition 26, 44, 119, 153, 154, 159, 229
Initial value problem 26, 29-52, 65, 66, 118, 120, 249-250, 264
Initial value technique 140
Initial-boundary problem 118, 153, 154
Inner boundary 220
Instability 46-52
Interior grid point 85, 99, 121, 230
Interior lattice point 85, 99, 121, 230
Interval of dependence 159
Inverse point 92
Inversion 92, 94, 101, 102
Inversion mapping 92, 94, 101, 102
Isentropic flow 246, 252
Isoperimetric inequalities 103, 105

Kinetic energy 266
Kutta's formulas 38-39, 40

Laplace difference analogue 80-83, 87
Laplace difference equation 80-83, 87
Laplace's equation 70, 73, 80, 98, 112
Lattice 85
Lattice point 83-86
Lax-Wendroff method 254-256

SUBJECT INDEX

Linear algebraic system 1-12, 88, 90, 131, 134, 148, 149
Linear differential equation 27, 57, 69, 95-98, 113
Linear partial differential equation 69, 95-98, 113
Liquid 208

Mach number 72
Machine error 46
Main diagonal 5, 60
Mathematical science 276-285
Matrix 1-12, 218, 241, 242
Max-min property 74, 82, 96, 126-127, 138
Maximization 186
Method D 86-91, 97, 99, 107, 112, 144, 181, 217, 237
Method of characteristics 182
Method of Courant, Isaacson and Rees 250-254
Method of Fromm 237-238
Method of Pearson 239
Method of Taylor series 29-32, 53
Mild diagonal dominance 3, 149, 176, 177, 179
Mildly nonlinear partial differential equation 70, 106-113, 138-150, 174-182
Minimal surface equation 71
Minimization 186, 203
Mixed type problem 79
Motion 259-260

Navier-Stokes equations 208-209, 239-240
Necessary conclusion 284
Neighbor 85, 87
Neumann problem 79
Newton's dynamical equation 43, 262
Newton's formula 15
Newton's method 15, 108
Nonlinear boundary value problem 63, 106-113

Nonlinear force field 262-264
Nonlinear pendulum 43-46, 185
Normal derivative 78-79, 103
Normal form 243, 247, 249, 253

Ordinary differential equation 26-66, 143, 180, 189
Oscillation 262-264
Over-relaxation factor 16
Overflow 46, 48, 50

Parabolic partial differential equation 70, 71, 118-150, 237, 239
Parameter 34, 80, 214
Partial differential equation 69, 212
  biharmonic 227
  elliptic 70-113, 212
  gas dynamical 71-72
  heat 71, 118-138
  hyperbolic 70, 71, 153-182
  Laplace 70, 73, 80, 98, 112
  linear 69, 95-98, 113
  mildly nonlinear 70, 106-113, 138-150, 174-182
  minimal surface 71
  Navier-Stokes 208-209, 239-240
  parabolic 70, 71, 118-150, 237, 239
  quasilinear 69, 241, 246, 249, 252
  soap film 7, 199
  wave 71, 153-182, 243
Particle 259-260
Particle-in-cell method 256
Pendulum 43-46, 263
Periodic function 56
Periodic solution **26**, 52-56
Physically reasonable 127
Piecewise regular 75
Plateau problem 198-204
Poisson integral 78

Potential energy 267
Potential equation 70
Primary vortex 226
Program error 46

Quasilinear partial differential
   equation 69, 241, 246, 249,
   252

Radiation equation 64
Rectangular integration 191
Reduced wave equation 182
Region of dependence 159, 164
Remainder 30
Reynolds number 209
Robin problem 79
Rounding 51, 126, 140
Row of grid points 121, 131,
   162, 179
Runge-Kutta method 32-43, 182

SOR 12, 62, 90, 91, 112
Saddle surface 74
Secondary vortex 211, 226
Semi-infinite strip 119, 154
Shortest path 192
Smoothing 224, 231, 239, 261
Soap film equation 7, 199
Sonic flow 72
Speed of sound 72, 246
Stability 121-128, 136, 139,
   164, 175, 238, 240, 252
Stable 127, 164, 168
Steady state 143, 145, 208, 229,
   234
Step-ahead technique 140
Stream function 209
Streamline 226, 234
String vibration 267-273
Subsonic flow 72
Successive over-relaxation 12,
   62, 90, 91, 112
Supersonic flow 72
Surface potential 103
Syllogism **284**

Symbol manipulation 29, 33
System of particles 267

Taylor expansion 29, 34, 36, 80,
   96, 142, 178
Taylor series 29-32, 35, 53, 214
Tension 268, 271
Three-point formula 28
Time 259-260
Total charge 103
Trailing wave 271
Transcendental system 1-22, 64
Trapezoidal integration 195
Triangularization 200
Tridiagonal 4, 5, 10-12, 58, 59,
   131, 134, 137, 139, 168, 169
Truncation 140
Two-point formula 28

UNIVAC 1108 5, 10, 45, 56, 104,
   150, 226, 233, 264
Undefined term 278-279
Unit circle 91
Unit cube 103
Unit sphere 101
Unstable 46

van der Pol's equation 52-56
Variational problem 185-186, 197
Velocity 261
Viscosity 208, 270
Viscous flow 208
Vortex 211, 226
Vorticity 209, 211

Wave equation 71, 153-182, 243
Work 264-266